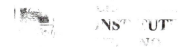

Designing Flexible Cashflows

Scott Fawcett

2003

A division of Reed Business Information
Estates Gazette, 151 Wardour Street, London W1F 8BN

ISBN 0 7282 0420 7

Typeset by Amy Boyle, Rochester, Kent
Printed in Great Britain by The Short Run Press Ltd, Exeter, Devon

Contents

Preface

If I had a penny for every time I've been asked "Can you help me with this simple cashflow problem" I'd probably have, at the very least, a whole bucket of pennies.

So, one weekend I decided to jot down a few brief instructions on how to set up a simple property investment cashflow. Well, I never did know when to stop and the result is the book you are now holding. I very much hope that it gives you some ideas that will improve your cashflow and analysis skills.

I'd like to take the opportunity to thank those of my colleagues at Drivers Jonas who have helped me in the preparation of this book. I'd also like to give special thanks to my wife, Claire, for her unfailing support and interest in this project, and to my parents, for buying me my first computer, a ZX81, and who are, therefore, really to blame for all this.

Scott Fawcett
August 2003

Chapter 1

Introduction

How flexible is flexible?

A cashflow is simply a list of numbers showing, in property investment terms, the rental income from an investment together with associated costs and capital expenditure, if required.

At its simplest, and quickest, typing in a series of numbers to show the income from an investment will do the job. However, the whole process becomes very time consuming if there are lots of tenants or properties, or if it is necessary to change your assumptions to see how variations impact the overall investment return.

Consequently, it is important to be able to write cashflows that can react to changing inputs. Flexible cashflows can incorporate any possible variation you can think of, for example variable ERVs (estimated rental values) and void periods, different rental growth rates for each element of the property, options to operate or ignore break clauses and so forth.

It is not hard to imagine that such a degree of flexibility is complex to design and is certainly not for the faint-hearted!

However, not all analyses will require this. It may suffice to simply ignore rental growth and be able to do no more than vary ERVs for a number of different tenants. Writing a cashflow to do this is fairly simple once you have learnt a few basics and will form a good platform to develop upon over time. Essentially, start small and build up.

There are plenty of examples contained within the following, and it is important to work through these and to try out your own variations. Given its widespread use, all of the examples use the Microsoft Excel®[1] spreadsheet and you will find the help files within this program of great assistance. At the end of the main text, there is a brief review of frequently used functions that should also provide a handy reference point.

The object is not to learn these methods by rote – rather to understand how the techniques work and how you can apply them. Every cashflow you write will be different and you will no doubt need to be able to apply these ideas in a whole host of innovative ways in order to solve the problems that will arise.

It is worth mentioning that the techniques described do not pretend to be the ultimate in cashflow design and it is recognised that you may already have your own preferred approach to dealing with some cashflow elements. However, if you do not like, for example, the way that void costs are treated, or growth is applied

---

[1] Microsoft® and Excel® are registered trademarks of Microsoft Corporation, but for convenience the text refers to this simply as "Excel". Screen grabs are reproduced by kind permission of the Microsoft Corporation.

then perhaps you can combine your own methods with some of the other ideas contained herein so as to improve the whole.

There are always ways to improve, simplify and create more elegant solutions. The challenge is to be able to evolve these yourself.

Chapter 2

Tools of the Trade

Essential functions that you need to know

If you are going to program a flexible cashflow, there are some functions that you will have to get to know and love. There are also techniques that you will need to grasp.

We will start with some of the Excel spreadsheet functions that will be of particular use.

The IF statement is your friend

The essence of a spreadsheet is to have the value of individual cells changing depending on the value of other cells. In a simple SUM calculation for example, the output of the formula will vary depending on the values in the SUM range but it will always give the summation of those cells.

The IF statement enables the spreadsheet author to do much more than this by allowing the operative formula (i.e. the formula which gives the cell's result) within the cell to be changed. So, taking the above example, the SUM formula could become a multiplication calculation.

Simple IF statements

At its simplest, the IF statement need not contain other formulae. The structure of the statement is as follows:

=IF(Logical Test, TRUE, FALSE)

The "Logical Test" is essentially a question and can use equals, greater than, less than, etc. comparators. If the answer to the question is true, then the value generated by the TRUE part of the function is used, if not then the FALSE value is used.

A simple example of how this works is as follows:

=IF(B10=5,"B10 equals 5","B10 does not equal 5")

The results of this are straightforward and, of course, only one number in B10 will generate the TRUE result.

It is worth briefly noting the use of the quotation marks. These are used to make text appear in cells, rather than figures. The above formula could have a numerical output if the following were used:

=IF(B10=5,200,1000)

Let's try using a simple IF statement.

1.   Open a blank spreadsheet and enter the following:

- B4: Input:
- B6: Is input 5?
- C4: 5
- C6: =IF(C4=5,"Yes","No")

2. Cell C6 should give the answer "Yes" while the number 5 is in C4.
3. Save this spreadsheet as we will use it in the next example.

	B	C
2	Simple IF Statement	
3		
4	Input:	5
5		
6	Is input 5?	Yes

4. Changing C4 from "5" will alter C6 to the false result.

IF and formulae

The next stage of using the IF statement is to introduce different formulae into the true and false outputs. Consider the earlier example, where an addition formula becomes a multiplication, this could be expressed as follows:

=IF(A1="Multiply", B1*B2*B3,SUM(B1:B3))
IF A1 equals "Multiply" then insert B1*B2*B3 otherwise add together B1, B2 and B3.

Here, the true part of the IF statement contains the multiplication and the false the addition. If A1 is anything other than the word "Multiply" then an addition will be carried out.

Note the quotes around "Multiply" to show a text input, the lack of the need for an equals sign before the SUM formula and the double brackets at the end of the formula – first closing the SUM formula and then the IF statement.

Here is an illustration of this in action.

1. Amend the previously saved spreadsheet by entering the following, over-writing cells where necessary:
   - B5: Input:
   - B6: Calculation:
   - B7: Result:
   - C5: 5
   - C6: Multiply
   - C7: =IF(C6="Multiply",C4*C5,Sum(C4:C5))
2. While C6 contains "Multiply", C7 will equal 25. Any other entry in this cell will make C7 equal 10.

	B	C
2	Simple IF Statement	
3		
4	Input:	5
5	Input:	5
6	Calculation:	Multiply
7	Result:	25

The problem with basic examples is that over-simplifying rather loses the point. However, you should be able to see from the above that not only can the inputs be changed to vary the result, but so can the type of calculation.

Nesting IF statements

It has been illustrated how an IF statement can make use of other functions. As an IF statement is a function, then an IF statement can contain another IF statement! This is known as "nesting" IF statements.

The new IF statement can take the place of either, or indeed both, the TRUE or FALSE arguments – and so this:

=IF(Logical Test, TRUE, FALSE)

Becomes one of the following, or a combination of the two:

=IF(Logical Test, IF(Logical Test, TRUE, FALSE), FALSE)
=IF(Logical Test, TRUE, IF(Logical Test, TRUE, FALSE))

We will leave the IF statement there for the moment. The next chapter will expand on the use of nested IF statements, but now we will look at how to broaden the logical questions that you can ask.

Logical functions

Named after the 19th-century mathematician George Boole, Boolean logic is a form of algebra in which all values are reduced to either a TRUE or FALSE result. They can be used to compare the results of several questions and generate a single answer.

TRUE and FALSE might not seem very useful outputs from these statements – but if you consider them in the light of how the IF statement works these are quite handy.

We will concentrate on the two most commonly used functions – "AND" and "OR". You may already be familiar with the application of these from their use in internet searches.

AND function

The AND function compares several statements and generates a TRUE result if, and only if, all of the individual statements are true. Consider the following:

A room contains two chairs and two people – Bill and George for the sake of argument. At any point either could be sitting or standing.

One could make the statement that "George and Bill are sitting". This would only be true if both were sitting. If either, or both, were not sitting then the statement would not be true.

In spreadsheet terms you could ask the same question – albeit structured slightly differently as we have to query what the two people are doing individually and we have to adopt the Excel spreadsheet's structure for the function.

The AND function is written as follows:

=AND(Logical 1, Logical 2, ... up to 30 arguments)
The logical arguments can use the =, <, >,<= and >= comparison operators.

So, the question would be expressed as:

=AND(George=Sitting, Bill=Sitting)

Why might you want to use this? A common application is when looking at rent reviews – for an upwards only-review to take place, the cashflow must be at the date of the review and the ERV must be greater than the passing rent, therefore:

=AND(Current Date=Review Date, ERV>Passing Rent)

Only if both of these are true should the reviewed rent be adopted.

It would be a worth having a go at using a simple AND statement at this point – just to get the feel of it. Try the following:

1. Open a blank spreadsheet and enter the following:
    *   B5: George
    *   B6: Bill
    *   B8: Both Sitting?
    *   C4: Sitting?
    *   C5: Y
    *   C6: Y
    *   C8: =AND(C5="y",C6="y")
2. Save the spreadsheet as we'll use it again in the next section.
3. Alter the C5 and C6 cells.
4. Anything other than a "Y" in both cells will result in a FALSE answer.

	B	C
2	AND Example	
3		
4		Sitting?
5	George	Y
6	Bill	Y
7		
8	Both Sitting?	TRUE

OR function

The OR function compares several statements and generates a TRUE answer if any of the logical statements within it are true.

We will return to our room and chairs and make another statement.

This time, we will say, "Either George or Bill are sitting". This would, of course, be true if either one or other were sitting – and also if both were sitting. Only if neither were sitting would the statement be false. You might feel that is stretching the language a little but that's life.

Again, in the spreadsheet the question would be structured slightly differently. Adopting the Excel spreadsheet's structure, the OR function is written in the same way as the AND statement:

=OR(Logical 1, Logical 2, ... up to 30 arguments)
The logical arguments can use the =, <, >,<= and >= comparison operators.

So, the question would be expressed as:

=OR(George=Sitting, Bill=Sitting)

To experiment, go back to the AND example that you created.

1. Amend the spreadsheet by over-writing cells as follows:
   - B8: Either Sitting?
   - C8: =OR(C5="y",C6="y")
2. Alter the C5 and C6 cells.
3. Anything other than a "Y" in one or both cells will result in a FALSE answer.

	B	C
2	OR Example	
3		
4		Sitting?
5	George	N
6	Bill	Y
7		
8	Either Sitting?	TRUE

A potential use of this might be to identify if your cashflow has reached one of a number of review dates (useful if you are using rental growth):

=OR(Current Date=Review 1, Current Date=Review 2)

If this were true, you could then ask whether ERV was greater than passing rent to decide whether to implement the review or not. To do this, you would combine this OR function with the previous AND example:

=AND(OR(Current Date=Review 1, Current Date=Review 2), ERV>Passing Rent)

LOOKUP functions

If you are feeling a little overwhelmed at this point but are keen to move on to creating the first cashflow, you might like to skip this section for the time being. You will be able to assemble the cashflow in the next chapter without using the following functions.

The lookup functions are not essential for cashflow work, but do help to automate some tasks, which helps to improve flexibility, and they will be used a great deal later on.

The function comes in three forms: HLOOKUP, VLOOKUP and LOOKUP. All are alike in their use, although the first two functions are particularly similar; the only difference being that one searches horizontally, the other vertically through an array.

HLOOKUP

This enables values to be sought across an array of data and the result is given from the same row or a row below the row searched. The format of the function is:

=HLOOKUP(Value, Range, Row, Range Lookup)

Value is the item to be searched for and this can be a number or text.

Range is the block to search and must include all of the rows that you wish to search down.

Row is the number of rows to look down once the Value is found. 1 will return a value from the same row that the search is performed on.

Range Lookup uses a logical value. If this is set to TRUE, or omitted, then an approximate (next largest) match will be returned if an exact match does not exist. If this is set to FALSE then only an exact match will be used. If there is no exact match then a "#N/A" error is given.

We can ask the spreadsheet to look for the letter "X" in the following array and return the figure from two rows below this:

	B	C	D	E	F
2	U	V	W	X	Y
3	1	2	3	4	5
4	100	200	300	400	500

This formula would be structured as follows:

=HLOOKUP("X",B2:F4,3)

Look for X in row B2:F2, once found return the value from the third row of the block.

Note that you only need quotes in the first part of the formula if you are searching for text. It is important to remember that the search is carried out in the top row and the range must include all of the rows that you wish to look down. These functions can sometimes be a little confusing and the last thing you need is a strange error occurring because you have tried to search outside the stated block.

There are various uses to which we will put this function. One of the more useful will be to work out lease restart dates following an expiry and a variable void period. This will be explained later.

VLOOKUP

This works in the same way as HLOOKUP – except rather than searching along a row for a value, it searches down a column and then moves across a specified number of columns to generate an answer.

LOOKUP

This function is similar to the others – but allows single, non-contiguous ranges to be used. This enables, for example, results above the search range to be found (remember that HLOOKUP only looks down a range).

There are two forms of this function, a "vector" and "array" form. We will use the vector form, which takes the following format:

=LOOKUP(Value, Lookup Range, Results Range)

Value is the item to be searched for and this can be a number or text.

Lookup Range is a single row or column.

Results Range is also a single row or column of the same size.

We could look for 400 in one block and select the corresponding letter as shown in the example below. The formula is as follows:

=LOOKUP(400,B4:F4,B2:F2)

	B	C	D	E	F
2	U	V	W	X	Y
3					
4	100	200	300	400	500

This function also has the advantage that it will continue to work if additional rows are inserted between the search and result ranges as Excel will automatically adjust the range references to allow for this.

If you inserted rows into a block referred to by an HLOOKUP, or similarly with columns in a VLOOKUP, expression you would have to amend the figure that tells the formula the number of rows to look down.

Chapter 3

Assembling a Simple Investment Cashflow

Putting theory into practice

Having spent some time introducing a few of the functions that you will need let's first give some consideration to the concept of how to structure a cashflow, before going on to build up an example piece by piece.

The cashflow concept

There are many ways to structure a cashflow. We will concentrate on one technique, which will introduce many of the basic elements. Once you have grasped the idea then you will no doubt evolve your own methodologies.

You will recall the structure of the IF Statement is:

=IF(Logical Test, TRUE, FALSE)

For the purposes of the following, it helps to translate this into something closer to English – especially when using nested statements. Generally, the form of nested IF statement that we will use will be:

=IF(Logical Test, TRUE, IF(Logical Test, TRUE, FALSE))

And it is best to think of it as:

Single Statement: **=IF(Question, Then, Otherwise)**
Nested Statement: **=IF(Question, Then, (Another Question, Then, Otherwise...))**

The key feature of the following method is to concentrate on picking out where rents change. If the rent does not change, then it must be the same as the quarter before, so just copy that.

Also, while it sounds awkward, an easy way to structure the formula is to start at the end and work backwards. Using the structure of the IF statement shown above, as soon as a TRUE answer is reached then that result is returned and the questioning stops.

The trick with this method is to be able to ask a series of questions such that, if the answers are all FALSE, the last action is to copy the previous cell. To illustrate, the structure will work as follows:

IF(Current Date = Lease Restart, then ERV, IF(Current Date = Lease Expiry, then 0, IF(Current Date = Rent Review AND ERV > Passing Rent, then Reviewed Rent, otherwise copy previous cell)))

Finally, to make the above work, it helps if all lease input dates are rounded to the nearest quarter day, as this enables us to keep the formulae as simple as possible.

We will build the above up stage-by-stage, but first, a gentle introduction to nesting IF statements.

Quarter days

Most people who need to assemble property investment cashflows have, at some point, tried to generate a list of quarter days. Often, they might end up adding 90 days or so to the previous quarter and ending up with:

25-Mar-03	25-Jun-03	25-Sep-03	26-Dec-03	27-Mar-04	27-Jun-04

After making adjustments every few years to get the sequence back in step (and formatting to take out the days figure) the result looks okay, but it is not accurate enough to compare our lease dates against.

Here is a way to generate accurate quarter days that will introduce you to a working example of a nested IF statement. It will also provide a useful base for generating cashflows later on.

First you need to work the logic of the problem out. We want this to be flexible, so we will be entering the quarter we want to start from and asking Excel to work out the rest. If March is our starting point, the argument works as follows:

- If the previous quarter was March, then the current date must be 24 June in the same year.
- If the previous quarter was June, then the current date must be 29 September in the same year.
- If the previous quarter was September, then the current date must be 25 December in the same year.
- If the previous quarter was December, then the current date must be 25 March in the next year.

To do this we will use the DATE, MONTH and YEAR functions. These are quite straightforward and so some simple examples will suffice. You can refer to the Help files in Excel for more detail if you need it:

=DATE(Year, Month, Day) generates a number that can be formatted as a date.
=MONTH(Cell) returns the month represented by the value in the cell.
=YEAR(Cell) works in the same way as the MONTH function.

The formula can be written as follows:

=IF(Month(Previous Cell)=3, DATE(Year(Previous Cell), 6, 24), IF(Month(Previous Cell)=6, DATE(Year(Previous Cell), 9, 29),IF(Month(Previous Cell)=9, DATE(Year (Previous Cell), 12, 25),IF(Month(Previous Cell)=12, DATE(Year(Previous Cell)+1, 3, 25)))))

Of course, this can be slightly simplified. In the above, the final FALSE statement is missing – after asking whether the previous month is September, there is no need for an extra IF statement – just use the FALSE part of the statement for the last calculation:

=IF(Month(Previous Cell)=3, DATE(Year(Previous Cell), 6, 24), IF(Month(Previous Cell)=6, DATE(Year(Previous Cell), 9, 29),IF(Month(Previous Cell)=9, DATE(Year (Previous Cell), 12, 25), DATE(Year(Previous Cell)+1, 3, 25))))

It is worth experimenting with this yourself as the technique is important. Do not feel that you have to tackle the whole formula in one go, here is how to build it up slowly:

1. Start by entering the initial quarter day, if you simply type "25–Mar–03" in A1 Excel will automatically treat this as a date and format it accordingly.
2. Then begin entering the formula in B1, but instead of entering the second IF statement, just put 999.
   - B1 = IF(MONTH(A1)=3,DATE(YEAR(A1),6,24),999)
3. By changing the date in A1 away from a March date the result in B1 will vary between 24–Jun–03 and 999.
4. Add the next IF statement by replacing the 999 with the test for a June date.
   - B1 = ...,24),IF(MONTH(A1)=6,DATE(YEAR(A1),9,29),999))
   - Note the extra ")" at the end of the formula to close the additional IF function.
   - Again, you can test this by changing the first date.
5. Repeat the above to add in the additional arguments.
6. Copy the formula across as far as you need it.
7. Save as "Quarter Day Calculation" or something similar. By simply copying A1 and B1 into other spreadsheets, and copying cell B1 along a row, you can add the quarter day generator to other cashflows with ease and you will need this frequently as we go on.

Just in case you have any trouble with this, as this formula will be used extensively, if your initial date is input in A1, your formula in B1 should be:

B1=IF(MONTH(A1)=3,DATE(YEAR(A1),6,24),IF(MONTH(A1)=6,DATE(YEAR(A1),9, 29),IF(MONTH(A1)=9,DATE(YEAR(A1),12,25),DATE(YEAR(A1)+1,3,25))))

A quicker approach

Having made you work through a rather lengthy example to get used to putting IF statements together, here is a different method using the EDATE function.

Before going any further, do make sure that the *"Analysis Toolpak"* is installed in your version of Excel by going to the Tools/Add-Ins menu and making sure that

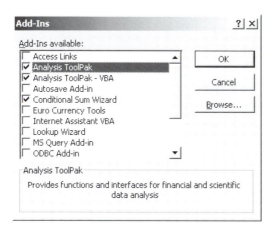

there is a tick in the box adjacent to "Analysis Toolpak". Some functions, EDATE included, will not be available without this feature installed. If there is no tick, click on the blank box to insert one.

The function is structured:

=EDATE(Start Date, Months)

It simply adds a number of months onto the start date – without changing the day figure. So, if you set up your first four quarter dates:

	B	C	D	E	F
2	25-Mar-03	24-Jun-03	29-Sep-03	25-Dec-03	

You can then use EDATE to add 12 months onto each quarter and extend the series:

F2 =EDATE(B2,12)
Add 12 months to the input date

	B	C	D	E	F
2	25-Mar-03	24-Jun-03	29-Sep-03	25-Dec-03	25-Mar-04

This formula can then be copied across the cashflow as far as required. In the following examples either method of generating dates can be used.

Building the cashflow

When it comes to actually putting a cashflow together, the first stage is to give some thought to what inputs you are likely to need – which will, of course, be determined by the results you need to generate.

For now, we will assemble a simple, yet flexible, cashflow allowing a single rent review to ERV but without including rental growth.

First, set up a tenancy schedule (note that the rent and ERV will be assumed to be on an annual basis in the schedule):

	B	C	D	E	F
2	Tenant	Rent	ERV	Review	Expiry
3	Smith	5,000	5,500	24-Jun-03	24-Jun-08
4	Jones	6,000	6,500	24-Jun-04	24-Jun-09
5	Brown	7,000	6,000	24-Jun-03	24-Jun-08

Also, paste the quarter day generator from the previous section into G2 and H2 (as you will need the start date and the calculation) and copy the date calculation across to generate dates up to 2010 or so (column AI).

	B	C	D	E	F	G	H	I
2	Tenant	Rent	ERV	Review	Expiry	25-Mar-03	24-Jun-03	29-Sep-03
3	Smith	5,000	5,500	24-Jun-03	24-Jun-08			
4	Jones	6,000	6,500	24-Jun-04	24-Jun-09			
5	Brown	7,000	6,000	24-Jun-03	24-Jun-08			

Normally, the schedule would include unit areas and a calculation of the ERV value from a square foot, or square metre, rate but these can always be added later.

Keep saving this cashflow as you add to it. As we go on we will be expanding its capabilities.

Starting at the end

We will start by simply picking out the expiry date. The formula will be entered into G3 and copied to the end of your dates. The logic for the formula is simple and you should be thinking along the lines of "**IF** the **current date** is **equal** to the **expiry date** then put in **zero**, otherwise do **nothing**". Highlighted are the elements of this that will go into your formula.

Try and write the formula yourself and test by copying along the length of the cashflow.

For the "do nothing" part, put "Next" into the formula – as this helps to highlight the parts of the cashflow that have yet to be addressed.

The formula should look like this (naturally you will not type in the G3 before the equals sign, this is only there to help you identify where the formula goes):

G3=IF(G$2=$F3,0,"Next")
IF Current Date = Expiry Date then 0 otherwise "Next"

Which, once copied across the cashflow, will generate the following, with the "Next" changing to 0 in June 2008 and nowhere else. Do not forget to fix the cell references to ensure that the calculation always compares the current date row against the expiry column (refer to the help files if you need detailed information on how to use absolute and relative cell references).

	B	C	D	E	F	G	H	I
						25-Mar-03	24-Jun-03	29-Sep-03
2	Tenant	Rent	ERV	Review	Expiry			
3	Smith	5,000	5,500	24-Jun-03	24-Jun-08	Next	Next	Next
4	Jones	6,000	6,500	24-Jun-04	24-Jun-09			
5	Brown	7,000	6,000	24-Jun-03	24-Jun-08			

A quick review

Now, we need to incorporate the rent review by replacing the "Next" part of the formula. This time, all we want to do is pick out the single point where the review is implemented – and we will assume, for the time being, that all reviews are upwards only.

Again, think through the logic that you need to apply – "**IF** the **current date** is **equal** to the **review date AND** the **ERV** is **greater than** the **passing rent**, then put in the **reviewed rent**, otherwise do **nothing**".

You may find it easiest to try this out on the second tenant to test it before amending the previous formula.

If you do this for the second tenant, you will need the following:

G4=IF(AND(G$2=$E4,$D4>$C4),$D4/4,"Next")
IF Current Date = Review Date AND ERV > Current Rent then insert the Quarterly ERV otherwise insert "Next"

You can copy this along to see what happens in June 2004 (note that columns are hidden between F and K in the next extract):

	B	C	D	E	F	K	L	M
2	Tenant	Rent	ERV	Review	Expiry	25-Mar-04	24-Jun-04	29-Sep-04
3	Smith	5,000	5,500	24-Jun-03	24-Jun-08	Next	Next	Next
4	Jones	6,000	6,500	24-Jun-04	24-Jun-09	Next	1,625	Next
5	Brown	7,000	6,000	24-Jun-03	24-Jun-08			

Try altering the review date and the ERV (changing the lease from reversionary to over-rented) to test the impact.

Now that you know how the next part of the formula works, you need to combine it with the first. Over-write the "Next" in the previous formula, in G3, as necessary. You should end up with:

G3=IF(G$2=$F3,0,IF(AND(G$2=$E3,$D3>$C3),$D3/4,"Next"))

IF Current Date = Expiry Date then 0, IF Current Date = Review Date AND ERV > Current Rent then Quarterly ER,V otherwise insert "Next"

Again, test the results. When you are happy that this works for the first tenant, copy it down for the other two, which will give the following result.

	B	C	D	E	F	G	H	I
2	Tenant	Rent	ERV	Review	Expiry	25-Mar-03	24-Jun-03	29-Sep-03
3	Smith	5,000	5,500	24-Jun-03	24-Jun-08	Next	1,375	Next
4	Jones	6,000	6,500	24-Jun-04	24-Jun-09	Next	Next	Next
5	Brown	7,000	6,000	24-Jun-03	24-Jun-08	Next	Next	Next

Fill in the gaps

Now, you have a cashflow that is reacting to the end date and the review. All you have to do is fill in the gaps!

We could add to, and complicate, the formula by querying whether the date is the start of the cashflow and, if so, inserting the current quarterly rent. Much simpler to over-write G3 and simply put in the initial rent. So, in G3 type:

G3=C3/4

And copy this down the spreadsheet:

	B	C	D	E	F	G	H	I
2	Tenant	Rent	ERV	Review	Expiry	25-Mar-03	24-Jun-03	29-Sep-03
3	Smith	5,000	5,500	24-Jun-03	24-Jun-08	1,250	1,375	Next
4	Jones	6,000	6,500	24-Jun-04	24-Jun-09	1,500	Next	Next
5	Brown	7,000	6,000	24-Jun-03	24-Jun-08	1,750	Next	Next

Finally, we need to remove the remaining gaps. As we have identified all of the changes in the cashflow, any remaining cells must remain unchanged, and therefore be the same as the cell before.

Amend H3 by replacing the "Next" part of the formula with "G3":

H3=IF(H$2=$F3,0,IF(AND(H$2=$E3,$D3>$C3),$D3/4,G3))
IF Current Date = Expiry Date then 0, IF Current Date = Review Date AND ERV > Current Rent then Quarterly ERV, otherwise copy previous cell

Now copy this down and across, to get:

	B	C	D	E	F	G	H	I
2	Tenant	Rent	ERV	Review	Expiry	25-Mar-03	24-Jun-03	29-Sep-03
3	Smith	5,000	5,500	24-Jun-03	24-Jun-08	1,250	1,375	1,375
4	Jones	6,000	6,500	24-Jun-04	24-Jun-09	1,500	1,500	1,500
5	Brown	7,000	6,000	24-Jun-03	24-Jun-08	1,750	1,750	1,750

There you have it – your first cashflow!

Chapter 4

More Complex Cashflows

Adding some additional features

Having assembled a basic cashflow with limited flexibility, let us move on a stage to look at how to incorporate variable void periods, new leases and break options.

Start all over again

The first additional elements we will look at are void periods and new leases. In the previous chapter the cashflow only dealt with a single lease so now we will add an extra lease following expiry.

This requires an extension to the tenancy schedule to allow an input for a void period and the calculation of a restart date, so insert new columns and labels as shown:

	B	C	D	E	F	G	H
2	Tenant	Rent	ERV	Review	Expiry	Void	Restart
3	Smith	5,000	5,500	24-Jun-03	24-Jun-08		
4	Jones	6,000	6,500	24-Jun-04	24-Jun-09		
5	Brown	7,000	6,000	24-Jun-03	24-Jun-08		

Working out the restart date

Calculating the restart date presents a similar problem to generating the quarter days. If we simply add three or six months to an expiry date we will miss the quarter day and the calculation will not react to the date (remembering that to keep things simple we always work to a quarter day).

Fortunately, we have already got a table of quarter days to use for reference, those that have already been generated to produce the cashflow, and so we can use HLOOKUP to select the appropriate quarter.

We need to search for a quarter closest to the expiry date plus a variable number of months void (we will use months, but you can adjust to quarters if you prefer).

First, put a figure in the void column, use six for the time being, and then give some thought to the structure of the formula and try to write this yourself. Remember that months are of different lengths and that if the formula does not find an exact match it will "drop back" and generate the highest preceding figure. So it is important that you make sure that the date you are looking up slightly exceeds the actual quarter day that is required.

You should end up with, in H3, something along the lines of:

H3=HLOOKUP(F3+G3*31+10,$I$2:$AM$2,1)
*Look for the Expiry Date + Void Months *31 + 10 in the Date Block and return the value in that row*

We have to add a few days onto the value of the expiry date and then search for that number in the list of quarter days. In the above, G3*31+10 roughly works out the number of days to add on to the expiry, by assuming 31 days in a month and then adds an extra 10 days on just to make sure that the result is slightly beyond the required quarter.

Remember that if your lookup range does not extend far enough, then the result will be the last figure in the range. If your formula does not seem to work do check this. When you create your own cashflows, it is usually worth having your quarters running far into the future to avoid any problems.

Once you are happy, copy down as required.

	B	C	D	E	F	G	H
2	Tenant	Rent	ERV	Review	Expiry	Void	Restart
3	Smith	5,000	5,500	24-Jun-03	24-Jun-08	6	25-Dec-08
4	Jones	6,000	6,500	24-Jun-04	24-Jun-09	6	25-Dec-09
5	Brown	7,000	6,000	24-Jun-03	24-Jun-08	6	25-Dec-08

It is worth testing this out against the range of void variation you are going to use, just to make sure that the "plus 10 days" approximation does not lead to any errors. If it does, you may need to slightly adjust this figure.

Naturally, there are ways of refining this but this method will get you started. You might like to give some thought as to how you could improve upon this.

Using the restart date

In the last chapter, we stopped the cashflow after the lease expiry. Now, we can extend it into a new lease.

This will require an addition to, and amendment of, the last formula. Previously, if the current date was after the expiry, then the cashflow should show zero. Now, if the cashflow is after the restart date it should show the ERV figure and only between the expiry and the restart dates should it be zero.

This is what you currently have in J3 (some of the references have shifted as we have added more columns):

J3=IF(J$2=$F3,0,IF(AND(J$2=$E3,$D3>$C3),$D3/4,I3))

Try to amend this to incorporate the restart; you will need another IF statement, and you should end up with:

J3=IF(J$2=$H3,$D3/4,IF(J$2=$F3,0,IF(AND(J$2=$E3,$D3>$C3),$D3/4,I3)))

IF the Current Date = Lease Restart, then insert ERV/4, IF Current Date = Lease Expiry, then insert 0, IF the Current Date = Review Date AND ERV > initial rent, then insert ERV, otherwise copy previous rent.

This can then be copied down to the other rows.

Take a break

Incorporating break options can be done in two main ways. You can either simply use the break as the expiry date – which is fine if you have a few tenants and are unlikely to want to examine the effects of operating different combinations of breaks. Or you can include a "switch" in the tenancy schedule to choose whether or not to operate the break.

We will look at how to implement the second option – it is good practice.

- First, you will need two more columns in your tenancy schedule – "Break Date" and "Operate", so insert them as shown.
- Then we need to amend the calculation of the restart date – based on either the break date or the expiry date depending on what choice you make in the schedule.
- Finally, we need to amend the cashflow to pick up the void on either the break or expiry.

	E	F	G	H	I	J
2	Review	Expiry	Break	Operate	Void	Restart
3	24-Jun-03	24-Jun-08			6	25-Dec-08
4	24-Jun-04	24-Jun-09			6	25-Dec-09
5	24-Jun-03	24-Jun-08			6	25-Dec-08

The "Operate" column will work from a Y or N input (although you will see that only the Y is really used) so put these into column H. It does not matter which goes where. Also, insert some break dates; any quarter day between the review date and the expiry will do.

Have a go at amending the restart calculation. All you need to do is add an IF statement to change the date calculation from being based on the expiry to the break as appropriate.

What you had in J3 was:

J3=HLOOKUP(F3+G3*31+10,$I$2:$AM$2,1)
Look for the Expiry Date plus the Void months * 31 + 10 in the quarter day block and return a result from the same row.

What you need to end up with is:

J3=HLOOKUP(IF(H3="Y",G3+I3*31+10,F3+I3*31+10),$K$2:$AM$2,1)
IF Operate = "Y" then lookup Break Date plus Void months * 31 +10, otherwise lookup Expiry plus Void months * 31 + 10.

Once you have copied this down, you can operate the break by putting a "Y" in the appropriate position (note that "Y" or "y" will both work here and in the formula). Anything else, you could leave it blank although here we have used "N", will not operate the break.

	E	F	G	H	I	J
2	Review	Expiry	Break	Operate	Void	Restart
3	24-Jun-03	24-Jun-08	24-Jun-05	Y	6	25-Dec-05
4	24-Jun-04	24-Jun-09	24-Jun-05	N	6	25-Dec-09
5	24-Jun-03	24-Jun-08	24-Jun-05	N	6	25-Dec-08

Now we need to amend the cashflow formula in a similar way. Your current formula in L3 reads:

L3=IF(L\$2=\$J3,\$D3/4,IF(L\$2=\$F3,0,IF(AND(L\$2=\$E3,\$D3>\$C3),\$D3/4,K3)))

IF Current Date = Restart Date then ERV/4, IF Current Date = Expiry Date then 0, IF Current Date = Review Date AND ERV > Current Rent the ERV/4 otherwise copy previous cell

Try amending the second IF statement **"IF(L\$2=\$F3..."** (which is "Does current date equal expiry date?") to read "**IF operate break equals "Y" then does current date equal break date, otherwise does current date equal expiry date?**".
The result will be:

L3=IF(L\$2=\$J3,\$D3/4,IF(IF(\$H3="y",L\$2=\$G3,L\$2=\$F3),0,IF(AND(L\$2=\$E3,\$D3>\$C3), \$D3/4,K3)))

IF Current Date = Restart then insert ERV/4, IF Current Date = Break Date (IF Operate is Y) or Expiry Date (IF Operate is not Y) then 0, IF Current Date = Review Date AND ERV > Current Rent then ERV/4 otherwise copy previous cell

Note the slightly unusual use of the third IF statement. Here, rather than supplying an answer, it supplies the question to the second IF statement.

Once copied over the cashflow you should have a calculation that will allow variable breaks for each tenant, restarts with ERV after a variable void period and implements rent reviews if there is a reversion.

You should also be feeling pretty pleased with yourself!

Chapter 5

The Addition of Rental Growth

This is where it gets a bit tricky!

Rental growth adds a whole host of issues, not least whether you need to adopt single or multiple growth rates. It is no longer appropriate to only consider a single rent review – each review in the lease needs to be considered and compared against the grown ERV. On lease renewal, a grown ERV needs to be adopted – and what about reviews in the renewed lease?

You should only tackle incorporating rental growth once you are completely happy with the previous chapters. The formulae can get quite complex and you will need to be comfortable with what you are doing.

How to implement growth

There are several ways to incorporate rental growth into a cashflow:

- You can create a rental growth index along the length of the cashflow (above the quarter days) to which you can refer. This has the advantage that you can easily vary the rate of growth over time.
- This can be made more flexible by having several growth indices along the cashflow – perhaps separate ones for retail, office and industrial elements.
- Or, you can use a different growth figure for each tenancy or property. Although this only allows the use of an average level throughout the cashflow, it does enable you to apply a wide range of individual growth rates.

The first and second of these are simplest as whenever you need an ERV figure you simply take the index figure in that quarter and apply it to your original ERV.

The third option makes formulae quite cumbersome, as you will need to calculate the growth figure every time you need to use it (thereby adding a $(1+i)^{\wedge n}$ type growth multiplier to each occurrence of the ERV figure in the formula).

To start, we will use the first method in the following and look at more complex rental growth in the next chapter.

On lease renewal

You will need to add some extra information into your tenancy schedule before we go much further. First, simplify the previous cashflow schedule by setting all of the operate break switches to "N" and hiding the Break and Operate columns – just to keep them out of the way.

The schedule now needs to include a note of all review dates – and you might as well add the rental growth index while you are about it. To do this, add an extra input for another review date, at column F, and then set up a growth index as follows.

Here, the "Growth Rate" heading is in G2 (you will need to insert an extra row to accommodate this) and the formula and inputs are:

K2=5%
L2=100
M2=L2*(1+$K$2)^0.25
Previous index figure multiplied by growth rate converted to a quarterly figure.

	B	C	D	E	F	G	J	K	L	M
2						Growth Rate		5%	100.0	101.2
3	Tenant	Rent	ERV	Review 1	Review 2	Expiry	Void	Restart	25-Mar-03	24-Jun-03
4	Smith	5,000	5,500	24-Jun-03	24-Jun-08	24-Jun-13	6	25-Dec-13	1,250	1,375
5	Jones	6,000	6,500	24-Jun-04	24-Jun-09	24-Jun-14	6	25-Dec-14	1,500	1,500
6	Brown	7,000	6,000	24-Jun-03	24-Jun-08	24-Jun-13	6	25-Dec-13	1,750	1,750

Then copy this across to column CQ. Note that, while setting up the cashflow, it's a good idea to use a high growth rate so that the changes in the cashflow are obvious.

In the schedule the lease expiry dates have been extended – so amend these. You will need to adjust the formula in your restart date calculation to cope with this change. To do this, first extend the quarter day generating formula to column CQ (this will give you plenty of scope to experiment with dates later on) and amend the formula in K4 to read:

K4=HLOOKUP(G4+J4*31+10,$L$3:$CQ$3,1)

Once that is done, and copied down the column, then adding the term to insert a grown ERV on lease renewal, rather than the original ERV, is relatively easy. Assuming that we are working in M4, you will need, in the appropriate part of the formula:

M4=IF(M$3=$K4,$D4*M$2/100/4,.......)
Where D4*M2/100/4 is ERV*Index/100 (to get the percentage increase)/4 (to convert to a quarterly rent)

Remember why we are now editing the formula in column M. Once growth is introduced, then at rent review we will need to compare a calculated ERV at the review date with the current rent payable. The easiest way to do this is simply to compare it with the rent payable in the previous quarter.

This introduces a problem – in the above if a review fell on 25 March 2003 then we would end up comparing an ERV figure against the Restart Date column.

A simple way to deal with this is to overwrite the first column with the current rent payable – be it the current quarterly rent or the current quarterly ERV (for example if a rent review was yet to be settled), so:

L4=$C4/4

If you have been editing the cashflow set up in the previous chapter this should already have been done.

If this seems like cheating, and if you want one formula that does everything, feel free to complicate matters by including within the formula another IF statement to put in a specific figure in the first quarter. It is all good practice!

Multiple reviews

That is the easy bit done, the next part is a little fiddly.

Now that the ERV is growing over time, we cannot simply compare ERV with the original rent to see if a review should take place. Every time there is a review we need to check how the current ERV compares with the immediately preceding rent.

We have already covered the OR statement that you will need for this element of the cashflow earlier on.

Where your formula tests for the review date and ERV, you need to ask "Is the current date equal to Review 1 or Review 2 and is the grown ERV greater than the passing rent?". Again, we will assume that we are working with upwards-only reviews.

We do this as follows (again assuming that we are working in M4):

M4 =...IF(AND(OR(M$3=$E4,M$3=$F4),$D4*M$2/100/4>L4), $D4*M$2/100/4,L4) ...

Which is equivalent to:

IF the Current Date=Review 1 OR the Current Date=Review 2 AND the Grown ERV is greater than the last rent, then insert Grown ERV, otherwise copy last rent.

This needs to be placed in your formula at the appropriate point. Once you have done that you should have a cashflow that fully incorporates rental growth.

All together now!

For the sake of simplicity the full formula has not been included above. So that you can see how it all fits together, the end result should be along the lines of:

M4=IF(M$3=$K4,$D4*M$2/100/4,IF(M$3=$G4,0,IF(AND(OR(M$3=$E4,M$3=$F4),$D4
*M$2/100/4>L4),$D4*M$2/100/4,L4)))
IF Current Date = Restart Date then insert grown quarterly ERV, IF Current Date = Expiry then 0, IF Current Date = either Review 1 or Review 2 AND Grown quarterly ERV is greater than the last quarter's rent then insert grown quarterly ERV otherwise copy previous cell

Based on the following tenancy schedule.

	B	C	D	E	F	G	J	K	L	M
2						Growth Rate		5%	100.0	101.2
3	Tenant	Rent	ERV	Review 1	Review 2	Expiry	Void	Restart	25-Mar-03	24-Jun-03
4	Smith	5,000	5,500	24-Jun-03	24-Jun-08	24-Jun-13	6	25-Dec-13	1,250	1,392
5	Jones	6,000	6,500	24-Jun-04	24-Jun-09	24-Jun-14	6	25-Dec-14	1,500	1,500
6	Brown	7,000	6,000	24-Jun-03	24-Jun-08	24-Jun-13	6	25-Dec-13	1,750	1,750

The cashflow generated takes the following form:

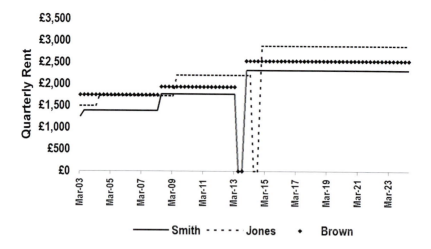

Make sure that you save a separate version of the cashflow at this point as we'll use it for a further example later on. Call it "Cashflow for Costs" for ease of reference.

Extending the cashflow

As can be seen from the graph, the income is fixed following the lease renewal. If you need to address this you can add additional reviews to the tenancy schedule and introduce another rent review date check as the first term of the cashflow, i.e. before checking for the expiry of the original lease, check for the review dates of the new lease.

Remember that you will need to link these rent reviews to the restart date, as they will have to vary depending on the changing void period.

One way to do this would be to use a similar method to the quarter day generator in order to extract the day and month of the review quarter and to add the review period onto the start year.

In other words you could set up new columns for your future reviews based on this type of formula:

=Date(Year(Restart Date)+Review Period, Month(Restart Date), DAY(Restart Date))

Another approach is considered below.

Rent reviews with MOD

The MOD function returns the remainder left after one number is divided by another. It takes the form:

=MOD(Number, Divisor)

So, MOD(3,2) asks, "What's left over if I divide three by two?" and the answer is, of course, one.

Why is this helpful? Well if, rather than looking at a single calculation, we consider a series, then a pattern emerges. Having set up periods 1–10 in row two in the example below, the formula in C3 is:

C3 =MOD(C$2, $B3)
Remainder once the Period number is divided by the Divisor.

This is copied down into B4 so that two examples can be compared. Using divisors of three and five results in the following series.

	B	C	D	E	F	G	H	I	J	K	L
2	Divisor	1	2	3	4	5	6	7	8	9	10
3	3	1	2	0	1	2	0	1	2	0	1
4	5	1	2	3	4	0	1	2	3	4	0

If the divisor is treated as being the rent review period, then it can be seen that the pattern repeats with the rent review frequency. Consequently, this can be used as a review trigger.

This could be used to generate reviews following a lease restart. Above, we considered calculating these by adding the review period to the restart date and setting up however many reviews we wanted. Using the MOD approach can provide regular reviews in perpetuity.

Here is an example of how you might set up the check for whether or not a particular quarter is a review quarter. As inputs, set up a Restart Date and a Review Period. Also, generate your quarter days.

The idea is first to consider whether you are in a review year, and then to check if the month is correct within that year.

The first part is:

D5 =MOD(YEAR(D2)-YEAR($B$5),$C$5)=0
MOD(Current Year – Restart Year divided by Review Period)=0

This will return a TRUE result in the review year:

	B	C	D	E	F	G	H	I	J	K
2	Restart	Review	29-Sep-09	25-Dec-09	25-Mar-10	24-Jun-10	29-Sep-10	25-Dec-10	25-Mar-11	24-Jun-11
3	Date	Period								
4										
5	25-Dec-05	5	FALSE	FALSE	TRUE	TRUE	TRUE	TRUE	FALSE	FALSE

This can then be combined with another test for the month:

D5 =AND(MOD(YEAR(D2)-YEAR($B$5),$C$5)=0,MONTH($B$5)=MONTH(D2))
AND(Year MOD test, Does the Current Month = Restart Month)

This will narrow down the range of TRUE results:

	B	C	D	E	F	G	H	I	J	K
2	Restart	Review	29-Sep-09	25-Dec-09	25-Mar-10	24-Jun-10	29-Sep-10	25-Dec-10	25-Mar-11	24-Jun-11
3	Date	Period								
4										
5	25-Dec-05	5	FALSE	FALSE	FALSE	FALSE	FALSE	TRUE	FALSE	FALSE

If you copy the formula across, you will see the TRUE result repeat every five years at the appropriate month. This could then be combined with other rent review tests, for example testing whether the reviewed rent is greater than passing rent, within your formula.

Chapter 6

More Rental Growth

Incorporating different rental growth rates

In the previous chapter we briefly considered different methods of incorporating rental growth. Here, we will look at two of these approaches in more detail so that you will be able to incorporate them if the need arises.

Variable rental growth

Sometimes, the use of a constant, average growth rate will not be appropriate and you will need to incorporate variable rates of growth.

Fortunately this is not difficult to do, all we need to do is insert different growth figures into the growth index at the appropriate points.

To do this, you will likely want to start with an input table, for example:

	B	C	D	E	F	G
2	2004	2005	2006	2007	2008	2009
3						& Beyond
4	4%	2%	0%	3%	5%	3%

Assume that our cashflow is going to run from 25 December 2003. The index at that point will be 100:

	B	C	D	E	F	G	H	I
2	2004	2005	2006	2007	2008	2009		
3						& Beyond		
4	4%	2%	0%	3%	5%	3%		
5								
6	Growth Index		100.00					
7	Quarter		25-Dec-03	25-Mar-04	24-Jun-04	29-Sep-04	25-Dec-04	25-Mar-05

The simplest way to vary the growth rate would be to adjust the formula each time the rate changes, thus:

E6=(1+$B$4)^0.25*D6
*(1+2004 Growth Rate)^0.25*Previous Index Figure*
I6=(1+$C$4)^0.25*H6
*(1+2005 Growth Rate)^0.25*Previous Index Figure*

Naturally, you would copy the E6 formula to F6, G6 and H6 before amending it in I6 and copying it further on.

There is nothing wrong with doing this – save it is a little tedious to input if you have more than a couple of years to do. Another approach would be to use the

LOOKUP function to match up the current year with the appropriate with growth rates:

E6=(1+LOOKUP(YEAR(E7),$B$2:$G$2,$B$4:$G$4))^0.25*D6

LOOKUP the current year in block B2:G2 and return the corresponding growth rate in B4:G4. Add 1 to this, raise to the power of 0.25 and multiply by the previous index figure.

You can now see why the layout of the input table looks slightly odd (with the "& Beyond" on the line below the years). If the LOOKUP function cannot find the years after 2009 in the table, it will just return the last value in the range – which is what we want.

Whichever route you choose, you should end up with:

	B	C	D	E	F	G	H	I
2	2004	2005	2006	2007	2008	2009		
3						& Beyond		
4	4%	2%	0%	3%	5%	3%		
5								
6	Growth Index		100.00	100.99	101.98	102.99	104.00	104.52
7	Quarter		25-Dec-03	25-Mar-04	24-Jun-04	29-Sep-04	25-Dec-04	25-Mar-05

Another benefit of using the LOOKUP approach is that you can now change the years that the growth rates apply to – as well as the rates themselves. In other words, changing 2009 in the table to 2010 would cause the 2008 growth rate to apply during 2009 as well. Handy if you need to quickly extend the length of variation.

Multiple indices

You may regard the use of a LOOKUP formula to set up the variable growth index as a little excessive. However, we can extend this idea and demonstrate why this is a useful technique.

You may want to be able to have more than one variable rate of growth. If you were creating a cashflow for a portfolio, this might be for properties with different uses or located in different towns.

First you will need your inputs, much the same as above but this time with several sets of information – in this case we will split these by use. You will also need room to work out your indices and a short form of cashflow inputs, just for the sake of example, such as:

	B	C	D	E	F	G	H
2		2004	2005	2006	& Beyond		
3	Industrial	1%	2%	3%			
4	Office	4%	5%	6%			
5	Retail	7%	8%	9%			
6							
7			Period	1	2	3	4
8							
9	Growth Index		Industrial	100.00			
10			Office	100.00			
11			Retail	100.00			
12							
13	Tenant	Use		25-Dec-03	25-Mar-04	24-Jun-04	29-Sep-04
14							
15	Smith	Office					
16	Jones	Retail					
17	Brown	Industrial					

Label the periods and quarters up to column AC (25 quarters). To insert the indices, use the same technique as previously – either manually insert the changes at the appropriate points or use the LOOKUP function (which should save you time):

F9=(1+LOOKUP(YEAR(F$13),$C$2:$E$2,$C3:$E3))^0.25*E9

LOOKUP the Current Year in the input block and return the corresponding number from the Industrial row. Add 1 to this, convert to a quarterly rate and multiply by the previous index figure.

Note that only the column references are fixed in the Industrial row reference – so that this will be updated as you copy the formula down to F10 and F11. These formulae can then be copied across the cashflow.

Now you need to pull the relevant index figure into your cashflow – ready for use as the growth term with your ERV figure. We have included a variable for this within the tenancy schedule inputs and we will link this in so that you are able to choose which index each ERV refers to.

We can do this as follows:

E15=VLOOKUP($C15,$D$9:$AC$11,E$7+1)

Search down column D in the indices block for the Use in C15 and then look across Period +1 columns and return this figure.

Copying this down for the three tenants and across will produce:

	B	C	D	E	F	G	H
2		2004	2005	2006	& Beyond		
3	Industrial	1%	2%	3%			
4	Office	4%	5%	6%			
5	Retail	7%	8%	9%			
6							
7			Period	1	2	3	4
8							
9	Growth Index		Industrial	100.00	100.25	100.50	100.75
10			Office	100.00	100.99	101.98	102.99
11			Retail	100.00	101.71	103.44	105.21
12							
13	Tenant	Use		25-Dec-03	25-Mar-04	24-Jun-04	29-Sep-04
14							
15	Smith	Office		100.00	100.99	101.98	102.99
16	Jones	Retail		100.00	101.71	103.44	105.21
17	Brown	Industrial		100.00	100.25	100.50	100.75

As the Use input is changed from "Office", "Retail" and "Industrial" so the cashflow will pick up a different index figure.

This is, of course, only a small element of the cashflow. To integrate this you will need to incorporate it each time you refer to ERV in the overall formula.

Individual growth rates

If you need to vary growth rates on a unit-by-unit basis you can adapt the formula used in the previous chapter. However, this does become slightly more complicated as instead of referring to the index calculation you will need to work out the growth factor for each row individually.

Let us start with a fresh working example based on work we have done so far. Set up a schedule as shown below. The cashflow will run from column L to CS and the only formulae you need to worry about are:

L3 and M3 – Insert the first quarter day and then the quarter day generator
L4=C4/4
M4=IF(M$3=$I4,$D4/4,IF(M$3=$G4,0,IF(AND(OR(M$3=$E4,M$3=$F4),$D4/4>L4),$D4/4,L4)))

IF Current Date=Restart Date then insert ERV/4, IF Current Date=Expiry Date then insert 0, IF Current Date= Review 1 OR Review 2 AND ERV/4 is greater than previous quarter's rent then insert ERV/4, otherwise copy previous cell.

Do not worry about setting up a formula for the restart date unless you want to – for the purpose of this example you can simply amend this by hand.

	B	C	D	E	F	G	H	I	J	K	L	M
2										Period	1	2
3	Tenant	Rent	ERV	Review 1	Review 2	Expiry	Void	Restart	Growth Rate		25-Mar-03	24-Jun-03
4	Smith	5,000	5,500	24-Jun-03	24-Jun-08	24-Jun-13	6	25-Dec-13	5%		1,250	1,375
5	Jones	6,000	6,500	24-Jun-04	24-Jun-09	24-Jun-14	6	25-Dec-14	3%		1,500	1,500
6	Brown	7,000	6,000	24-Jun-03	24-Jun-08	24-Jun-13	6	25-Dec-13	5%		1,750	1,750

You now have a basic spreadsheet that will insert ERV into the cashflow at the appropriate points, but does not incorporate any further rental growth.

The formula needs to be adapted to calculate the grown ERV on renewal and at review dates. That means we have to introduce the following growth term to replace the simple ERV * Index calculation. Working in cell M4:

Growth Term = (1+$J4)^((M$2-1)/4)
Growth Term = (1+Growth Rate)^(Current Period -1)/4)

In the above, 1 is deducted from the current period as period numbering starts from 1 and the ERV is as at the first period. Therefore, in period 2 only one period of growth is required.

This figure is then divided by four, as the growth rate needs to be on a quarterly basis. You may prefer to number your periods as 0, 0.25, 0.5, 0.75, 1... to simplify this.

Now, all you need to do is multiply the ERV by this figure each time it appears in the formula:

M4=IF(M$3=$I4,$D4*(1+$J4)^((M$2-1)/4)/4,IF(M$3=$G4,0, IF(AND(OR(M$3=$E4, M$3=$F4),$D4*(1+$J4)^((M$2-1)/4)/4>L4),$D4*(1+$J4)^((M$2-1)/4)/4,L4)))

*IF Current Period = Restart Date, ERV*Growth Term, IF Current Period = Expiry Date, then 0, IF Current Date = either Review Date AND ERV*Growth Term > Previous Quarter's Rent, then ERV * Growth Term, otherwise Previous Quarter's Rent*

Perhaps the easiest way to enter this is to build up the formula with "ERV" at each appropriate point (i.e. at the restart date, the review dates and in the review test) in the same way as we used "Next" in the first cashflow. Having written the ERV*Growth term expression once you can simply copy and paste this over the other "ERV" occurrences.

Chapter 7

Further Complexities

You have to know when to stop!

Given that you have stuck with it this far you can probably put together a quite complex cashflow incorporating rental growth, break options, lease restarts, etc. You are also probably starting to think of a whole host of complications that come up in leases and thinking of ways around them.

We will consider a couple of these below, but before that a brief word on the subject of knowing when to stop.

It is easy to over-complicate a cashflow. Imagine that you are setting up a cashflow for an industrial estate and you have one unit that has a stepped rental increase. You could add another term to the formula you completed in the previous chapter to allow for this and apply it to all tenancies. Or, you could simply over-write the formula at the appropriate quarter in this one instance with the rent you need.

Okay, so that is not very flexible – if for any reason you need to change the main formula you will have to be careful not to erase your manual amendment – but it is a very practical way to deal with a small variation.

It is up to you to gauge the required trade-off between flexibility and efficiency.

If you decide that the manual amendment route is for you, do remember to highlight where you have made the amendment (use a red font perhaps) or make use of the comment function within Excel (select the cell and use the Insert/ Comment menu).

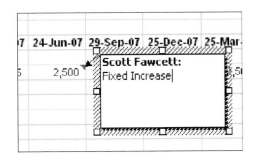

If you want your life to become more complex then try the following.

Vacant units

If you have vacant units to deal with, you will probably want to interpret elements of the tenancy schedule in a slightly different way – as dates you enter for reviews and expiries will be what you expect to get from the first letting. You will also need to incorporate an input column for a lease start date. But why go to all of this trouble?

Imagine that you have 10 vacant units on an industrial estate. How long a void should you allow before each unit is let – three, six, 12 months? The only thing you can be sure of is that somebody will have a different view to you and will ask you to change your assumptions.

The last thing you want to have to do is change several sets of review and expiry dates just because you are asked what difference a 12-month void makes rather than a six-month void, only then to be asked what happens if you let one unit each quarter.

So, you will want to design a tenancy schedule that does all the hard work for you. Start with the following headings and inputs:

	B	C	D	E	F	G	H	I
2		Term	Review		Rent Review			
3	Start Date	(Years)	Period	First	Second	Third	Fourth	Expiry
4								
5	29-Sep-03	15	5					

The easiest figure to calculate is the expiry date (and we will need this for use in the following stages). Using the DATE function in I5:

I5=DATE(YEAR(B5)+C5,MONTH(B5),DAY(B5))
DATE(Start Year + Term, Start Month, Start Day)

You use a similar technique to calculate the review dates – but of course we need to limit these to being earlier than the expiry date.

We will look at two ways to do this – just to give you some ideas.

Review dates using MIN

The calculation needs to add the rent review period onto the lease start date, but to stop once it reaches the expiry date.

We can do this using the MIN function, in E5:

E5=MIN(DATE(YEAR($B5)+$D5,MONTH($B5),DAY($B5)),$I5)
Return the lesser of Start Date plus Review Period and Lease Expiry Date

And then, using a similar formula, in F5 which can be copied across into G and H, but this time making sure that you add the review period to the previously calculated date, rather than the start date:

=MIN(DATE(YEAR(E5)+$D5,MONTH(E5),DAY(E5)),$I5)
Return the lesser of the previous Review Date plus the Review Period and the Expiry Date

This will give you the following:

	B	C	D	E	F	G	H	I
2		Term	Review		Rent Review			
3	Start Date	(Years)	Period	First	Second	Third	Fourth	Expiry
4								
5	29-Sep-03	15	5	29/09/2008	29/09/2013	29/09/2018	29/09/2018	29/09/2018

Any review dates that are greater than the expiry date will not be relevant and the expiry date will be inserted. If you are using the formula structure from the last chapter this will not matter as the expiry date is checked before the review dates.

So, effectively, the third and fourth reviews will be ignored – but it is not very elegant.

Review dates using IF

The calculation can be done with IF instead, in order to tidy things up. Again in E5:

E5=IF(DATE(YEAR($B5)+$D5,MONTH($B5),DAY($B5))>=$I5,"",DATE(YEAR($B5)+$ D$5,MONTH($B5),DAY($B5)))

IF the Start Date plus the Review Period is greater than or equal to the Expiry Date insert a blank otherwise insert the Start Date plus the Review Period

This may seem a bit unwieldy – but it is very simple to write. As the DATE calculation is repeated, get this working first and then copy it for use in your IF statement.

In the next column, just as in the MIN example, you need to add the review period onto the last review date. The only refinement you need is to stop the formula trying to work out a date from a blank cell – which will generate an error:

F5=IF(E5="","",IF(DATE(YEAR(E5)+$D5,MONTH(E5),DAY(E5))>=$I5,"",DATE (YEAR(E5)+$D5,MONTH(E5),DAY(E5))))

IF the previous cell is blank, return a blank, otherwise IF the previous Review Date plus the Review Period is greater than the Expiry Date return a blank, otherwise insert the Previous Review Date plus the Review Period

Once copied to columns G and H, this will generate a neater result than before:

	B	C	D	E	F	G	H	I
2		Term	Review		Rent Review			
3	Start Date	(Years)	Period	First	Second	Third	Fourth	Expiry
4								
5	29-Sep-03	15	5	29/09/2008	29/09/2013			29/09/2018

There are, of course, other ways you can do this – rather than label the Review columns "First, Second, etc." labelling them "1, 2, 3, 4" would enable you to multiply the review period by the number of the review and add that to the Start Date for example.

You are sure to evolve your own refinements.

Generating the cashflow

Once the tenancy schedule is set up, you can approach the cashflow in exactly the same way as we have done previously. The only real change is that you will need to add an additional term to check whether or not you have reached the start date and, if so, insert the ERV for the property – applying growth if necessary.

You can either do this by adding another IF statement after the formula has checked whether or not you are at a review date, or you can wrap the whole formula in an IF statement asking the same question.

Stepped rents

Considering the example mentioned in the introduction to this chapter: what if you have many leases that include stepped rents? How can you get the cashflow to incorporate these?

Often, the problem might be too complex. Stepped rents throughout a multi-let property are unlikely to fall in such perfect unison that they can all be accommodated into a few spreadsheet columns. However, if the range of dates is not too broad, perhaps you can round them to only a handful of key quarter dates, then there is a way to accommodate these.

Consider the following, we will keep everything in the schedule simple in order to concentrate on the stepped rents:

	B	C	D	E	F	G	H	I
2		Initial	Stepped Increases				Rent	Lease
3	Tenant	Rent	24-Jun-05	25-Dec-05	24-Jun-06	ERV	Review	Expiry
4								
5	Jones	£5,000	£5,500	£6,000	£6,500	£10,000	24-Jun-07	24-Jun-12

Looking at how this is set up, your first instinct might be that one of the lookup functions would be appropriate – and you could use these but this route will generate problems such as dealing with errors caused by searching the date block in D3:F3 for values less than the earliest date and so forth.

However, a solution can be devised using the SUMIF function, which takes the following form:

=SUMIF(Range, Criteria, Sum Range)
Range is the block of cells to be evaluated.
Criteria is what to look for – it can be numerical, an expression or text.
Sum Range is the corresponding block of cells to sum. If omitted, the Range block is used.

By way of an example as to how this function works, it could be used to show how a portfolio value is split between various uses:

	B	C	D
2	Property	Use	Value
3	A	Retail	£100,000
4	B	Office	£350,000
5	C	Retail	£250,000
6	D	Office	£500,000
7	E	Office	£1,000,000
8			
9	Total		£2,200,000
10			
11	Split by Use	Office	£1,850,000
12		Retail	£350,000

The formula in D11 is:

D11=SUMIF($C$3:$C$7,C11,$D$3:$D$7)
SUMIF(Use range, Office, Value range)

Copied down into D12, the calculation will operate on the Retail use.

Returning to the stepped rent example, rather than using the function for addition, we can use it to pick out a single number. In the spreadsheet below, B10 contains:

B10=SUMIF($D$3:$F$3,B8,$D$5:$F$5)
SUMIF(Stepped Increase Dates, Current Date, Stepped Increase Amounts) – i.e. add together all of the stepped increase amounts corresponding to the current date.

Once copied across the cashflow, this has the effect of inserting the stepped rent amounts at the appropriate points (ignoring the fact that we need to convert these to a quarterly rate for the moment).

	B	C	D	E	F	G	H	I
2		Initial	Stepped Increases				Rent	Lease
3	Tenant	Rent	24-Jun-05	25-Dec-05	24-Jun-06	ERV	Review	Expiry
4								
5	Jones	£5,000	£5,500	£6,000	£6,500	£10,000	24-Jun-07	24-Jun-12
6								
7								
8	25-Dec-04	25-Mar-05	24-Jun-05	29-Sep-05	25-Dec-05	25-Mar-06	24-Jun-06	29-Sep-06
9								
10	-	-	5,500	-	6,000	-	6,500	-

If you wanted to extend this to cover more stepped rents, all you would need to do is extend the blocks that the formula refers to – another tenancy could be incorporated by adding additional columns:

	B	C	D	E	F	G	H	I	J	K
2		Initial	Stepped Increases						Rent	Lease
3	Tenant	Rent	24-Jun-05	29-Sep-05	25-Dec-05	25-Mar-06	24-Jun-06	ERV	Review	Expiry
4										
5	Jones	£5,000	£5,500		£6,000		£6,500	£10,000	24-Jun-07	24-Jun-12
6	Smith	£9,000		£10,000		£12,000				
7										
8	25-Dec-04	25-Mar-05	24-Jun-05	29-Sep-05	25-Dec-05	25-Mar-06	24-Jun-06	29-Sep-06	25-Dec-06	25-Mar-07

However, if you were using this approach, just looking at the "Jones" cashflow line, you would need to ensure that in the event that a cell is blank the formula does not interpret this as a fixed "increase" to zero (unless that is what you intend). So, to incorporate this into the overall cashflow you could do the following:

B10=C5/4
The usual trick to get the initial rent into the cashflow that enables us to simply copy the previous cell in the next formula.
C10=IF(C8=$K$5,0,IF(AND(C8=$J$5,$I$5/4>B10),$I$5/4,IF(SUMIF($D$3:$H$3,C8,$D$5:$H$5)>0,SUMIF($D$3:$H$3,C8,$D$5:$H$5)/4,B10)))

IF Current Date = Expiry Date then insert 0, IF Current Date = Review Date and ERV is >
previous quarter's rent then insert ERV/4, IF the SUMIF statement is > 0, then insert relevant
stepped rent/4, otherwise copy the previous cell.

The result of which is that blank cells within the stepped rent block are ignored. This means that you could include a very comprehensive table for the inclusion of stepped rents with relative ease. If you have more than a few columns though you may want to hide these for presentation purposes.

	B	C	D	E	F	G	H	I	J	K
2		Initial			Stepped Increases				Rent	Lease
3	Tenant	Rent	24-Jun-05	29-Sep-05	25-Dec-05	25-Mar-06	24-Jun-06	ERV	Review	Expiry
4										
5	Jones	£5,000	£5,500		£6,000		£6,500	£10,000	24-Jun-07	24-Jun-12
6	Smith	£9,000		£10,000		£12,000				
7										
8	25-Dec-04	25-Mar-05	24-Jun-05	29-Sep-05	25-Dec-05	25-Mar-06	24-Jun-06	29-Sep-06	25-Dec-06	25-Mar-07
9										
10	1,250	1,250	1,375	1,375	1,500	1,500	1,625	1,625	1,625	1,625

Chapter 8

The Best Things In Life Are Free...

...but in a cashflow, you'll have to pay!

Unfortunately, in many investment scenarios it is not all just money in – there are often costs associated with running an investment property. These vary widely, but common examples that you will come across are professional fees, maintenance expenditure and other void costs.

Costs can be incurred in a wide variety of ways and we cannot hope to address them all. However, the following should give you some clues as to how to deal with a few common problems and give you ideas for how other charges can be dealt with.

Overall concept

In most cases, the easiest way to incorporate costs is to set up a cost cashflow below the income cashflow that has a line available for each unit in the property (or property in the portfolio). This way, you can see clearly when costs are incurred and keep the calculations fairly simple.

As the maths is generally quite straightforward, this is a quick way to set things up. It is also easy to add more than one cost cashflow if you really want to separate things out – perhaps one for professional fees and another for void costs.

Having constructed this, you can either total up all of the costs and income to give an overall net cashflow or examine the cashflow on a net income by tenant basis with ease.

Professional fees

The most common professional fees tend to be those on lettings and rent reviews. The cost calculations can either be linked to the tenancy schedule that you have input – which includes all of the crucial lease dates – or to changes in the cashflow. Frequently it might be a combination of the two.

So, we will write a formula to insert a letting fee after the lease expiry and to allow for review fees when there is a rental uplift.

Letting fees

Using the rental growth cashflow from the previous chapter, as there is only one point where a letting fee is due, which could be confused with a rent review event (you will see why below), we will deal with this first.

The letting fee will be due on the restart date for the new lease. We know when this will be, but we do not know what the new rent is. Rather than duplicating work, we can lift this from the income cashflow.

Set up the following based on the previously saved "Cashflow for Costs" spreadsheet. The only reason for using this is to give a varied income flow to base the costs on, if you have had any problems, just make something up!

The only formula which has been added is a simple SUM statement in L8 and beyond, the rest is just typed in.

	B	C	D	E	F	G	J	K	L	M
2						Growth Rate		5%	100.0	101.2
3	Tenant	Rent	ERV	Review 1	Review 2	Expiry	Void	Restart	25-Mar-03	24-Jun-03
4	Smith	5,000	5,500	24-Jun-03	24-Jun-08	24-Jun-13	6	25-Dec-13	1,250	1,392
5	Jones	6,000	6,500	24-Jun-04	24-Jun-09	24-Jun-14	8	25-Dec-14	1,500	1,500
6	Brown	7,000	6,000	24-Jun-03	24-Jun-08	24-Jun-13	6	25-Dec-13	1,750	1,750
7										
8	Gross Income								4,500	4,642
9										
10	Professional Costs									
11										
12	Relettings		Percent of New Rent		10%					
13										
14										
15										
16										
17	Smith									
18	Jones									
19	Brown									

The costs formula starts by testing for the Restart Date and, if appropriate, inserting the reletting fee:

L17=IF(L$3=$K4,L4*4*$F$12,"Next")

*IF Current Date = Restart Date then quarterly rent *4 *Percent of New Rent, otherwise "Next"*

So, if the Restart Date is reached, then the current quarter's rent is multiplied by 4 (to get the annual figure) and 10% of this is charged. In all other cases, the word "Next" appears.

Rent review fees

Rent review fees can be charged in many different ways. They could be based on a percentage of the uplift achieved or a percentage of total new rent. They may be subject to minimum figures – whether there is an increase or not.

For this example we will use a percentage of the new rent, subject to a minimum charge. A fee will only be payable if an uplift is achieved.

How best to identify when the charge should be levied? With a rent review it is pretty clear-cut: if the rent has gone up, a fee is due (provided it is not a lease renewal, which is the potential confusion mentioned above). We could link the test to the review dates in the tenancy schedule, but for the moment we will just examine changes in the cashflow.

The question you need to ask is "**IF** the **current rent** is greater then the **previous quarter's rent** then insert the higher of the **minimum fee** or the **percentage of the new fee**".

To do this, we will use the MAX formula, which, like its counterpart MIN, can simplify all kinds of applications. Here, we need to insert the higher of a fixed fee or a variable fee. You could do this with an IF statement:

=IF(X>Y,X,Y)

But imagine if you had more than two options – it would start getting quite complex so it is worth being aware of this function.

MAX returns the highest of a selection of values, the format is:

=MAX(Value 1, Value 2,........,Up to 30 Values)

Here, we will use:

=MAX(Variable Fee, Fixed Fee)

First, add in the inputs for the review fee:

	B	C	D	E	F	G	J	K	L	M
2						Growth Rate		5%	100.0	101.2
3	Tenant	Rent	ERV	Review 1	Review 2	Expiry	Void	Restart	25-Mar-03	24-Jun-03
4	Smith	5,000	5,500	24-Jun-03	24-Jun-08	24-Jun-13	6	25-Dec-13	1,250	1,392
5	Jones	6,000	6,500	24-Jun-04	24-Jun-09	24-Jun-14	6	25-Dec-14	1,500	1,500
6	Brown	7,000	6,000	24-Jun-03	24-Jun-08	24-Jun-13	6	25-Dec-13	1,750	1,750
7										
8	Gross Income								4,500	4,642
9										
10	Professional Costs									
11										
12	Relettings		Percent of New Rent		10%					
13										
14	Rent Review Fees		Minimum		500					
15			Percent of New Rent		10%					
16										
17	Smith								-	557
18	Jones								-	-
19	Brown								-	-

The formula that you need is:

L17=IF(L4>K4,MAX(L4*4*$F$15,$F$14),0)
IF current quarter's rent is greater than last quarter's rent, then insert the higher of the current rent multiplied by four times the Percent of New Rent and the Minimum Fee.

This needs to be inserted in place of "Next" in the letting cost formula in L17, so you end up with:

L17=IF(L$3=$K4,L4*4*$F$12,IF(L4>K4,MAX(L4*4*$F$15,$F$14),0))

While this works in the above example, you will probably notice that if the rent in L4 were greater than 41,633 (the numeric equivalent of 25–Dec–13) then the formula would treat this apparent increase as a review. You could close this loophole in several ways and you might like to think about how you would do this (the simplest method being to over-write the offending formula).

Going forward

What you need to develop from the above is knowledge of a variety of methods of pinpointing where costs arise:

- Above we have used an increase in passing rent to trigger a cost.
- At other times you might want to distinguish between a rise in rent from an existing rent (for a review) and from £0 (in the case of a reletting).
- Sometimes you will need to allow for the presence of fixed increases in rents, which should not trigger a review.

Thinking of ways around these types of problem is good practice for general problem solving in cashflows.

Void costs

Other common costs that you will need to deal with are those that arise due to vacancies – for example irrecoverable service charges and rates.

In most modern, full repairing and insuring leases these costs will only be relevant while units are vacant, hence the title, and we will make this assumption to be going on with.

They differ from the professional costs as, rather than being charged on rent, they usually relate to the amount of space occupied. So far, we have left the unit areas off the tenancy schedule, so as to keep things to a minimum, but now we will need to incorporate them – either in the original tenancy schedule or within the costs calculation area. For ease at this stage we will use the latter option.

Set up your inputs for void costs below the professional costs. One way to do this is to use a table of costs culminating in a single total. That way, you can see all of the inputs at a glance and extend the table if necessary.

	B	C	D	E	F
21	Void Costs				
22			Service		
23		Area	Charge	Rates	Total
24					
25	Smith	1,100	£2.50	£2.00	£4,950
26	Jones	1,300	£2.50	£2.00	£5,850
27	Brown	1,200	£2.50	£2.00	£5,400

In the above F25 contains:

F25 =(D25+E25)*C25

To insert the costs into the cashflow, all you need to do, for each tenant, is to pick out the points in the income cashflow where no income is received. So, in this case, L25 is:

L25=IF(L4=0,$F25/4,0)
IF income in current quarter is 0, then insert the annual void cost divided by four, otherwise insert 0.

Copy this across and down and costs should appear wherever there is a void.

Further complications

Again, you should think around what problems you might face and how you would solve them:

- How would you deal with a rates "holiday"?
- What if part of the void period is actually a rent-free period (so no rent is payable but the tenant is liable for service charges and rates)?
- How would you go about indexing the costs if you wanted to grow them over time?

Again, these are all good problems to wrestle with – better to consider these things while you have time than when you are being pressed to produce answers!

Chapter 9

Using Other People's Money

How to add finance to the equation

At the heart of many property transactions is the use of debt. Using bank debt can greatly enhance investment returns and it is important to be able to incorporate the impact of this into a cashflow.

The way to a truly flexible finance cashflow is to make sure that at any given point in the cashflow you know what interest and capital repayment you need to make, and how much debt is outstanding. By doing that, whenever you want to stop the cashflow, you can pick up the amount of outstanding debt that must be deducted from any sale proceeds.

Fortunately, the Excel spreadsheet has several formulae to help simplify this, which we will consider below. First though, it is important to understand how to convert interest rates between different periods and to understand how amortisation works.

Interest rate conversion

You must ensure that the interest rate you adopt is appropriate for the repayment frequency you are using – in other words, if you have an annual interest rate you will need to convert it to an appropriate periodic rate to fit in with your cashflow.

There are two simple ways to do this:

- You can either simply divide the rate by four – which implies that the annual rate is a nominal rate.
- Or, you can convert the rate using:

  Periodic Rate = $(1+i)^{(1/n)}-1$
  Where:
  i is the annual rate; and
  n represents the number of periods to spread it over

The second method is used if the annual rate is the effective rate.

To illustrate the difference, taking 10% pa would result in 2.5% per quarter using the first method (which compounded quarterly is equivalent to 10.38% pa) or 2.41% per quarter using the second (which compounded quarterly would equal 10% pa).

Either method might be appropriate depending on individual circumstances and you will need to give thought to which should be adopted in each case.

Loan amortisation

Often, you will want to establish how much your loan is going to cost on a periodic basis – perhaps so that you can deduct this amount from your income.

To use some of the financial functions, you will need to ensure that the "Analysis Toolpak" is installed by going to the Tools/Add-Ins menu and making sure that there is a tick in the box adjacent to "Analysis Toolpak" as some functions will not work if this feature is not installed.

In Excel, the formula to calculate a loan repayment is:

=PMT(Rate, Number of Payments, Present Value, Future Value, Type)

The components of the formula are used as follows:

- **Rate** is the interest rate for the loan. Remember, this must be on the same basis as the periodic payment. For example, you will need to convert your rate to a quarterly amount if using a quarterly cashflow.
- **Number of payments** is, not surprisingly, the total number of payments for the loan.
- **Present value** is the amount of the loan.
- **Future value** is the amount to which the loan is to be amortised to. If you are treating the loan amount as a positive number, this figure must be negative.
- **Type** allows you to choose payments in advance (value is 1) or arrears (value is 0 or omitted).

The output from the formula is a figure that will amortise the loan down to the specified amount over a given period. However, it is difficult to know intuitively whether or not the answer is correct – so it is worth working through an example to illustrate how the maths works and so ensure that you are comfortable that the formula works as you expect it to before we go on to simplify matters.

Let us imagine that you are going borrow £90,000. You want to calculate what the quarterly repayments will be under various scenarios and want to be able to vary the amount to which the loan is amortised to.

We can set up a loan amortisation schedule to model this (do refer to the example below to help set this up).

1. Start by considering the variables that you will need. Set up inputs for the loan amount, the interest rate, period of the loan (which we do not need just yet, but will need it later on), the sum to amortise to and the amount of the periodic repayment.
2. As described above, somewhere (either in a separate cell or within the repayment formula) you will need to ensure that your interest rate matches your repayment frequency. In the following example, we have simply used 2.5% per period.
3. Set up a table with the current period, opening balance, periodic payment, interest charge, capital repaid and an end balance as column headings.
4. Number the periods 1–8.
5. The opening balance will equal the loan amount – in the example C13=D4.
6. For the time being, pick any figure for the periodic repayment – say £10,000. This goes in D8 and D13=$D$8.
7. The idea is that you calculate the interest due in each period and any excess from the periodic payment goes towards repaying capital. So, in the first period, the interest repaid is the periodic interest rate multiplied by the opening balance (E13=$C13*$D$6).

8. The capital repaid is the excess. So, F13=$D13-$E13.
9. The end balance is simply the opening balance less the capital repaid. Here, G13=$C13-$F13.
10. In the second period, the opening balance is equal to the previous period's end balance (C14=G13).
11. The formulae are copied down throughout the schedule.
12. Note the inputs for "Number of Periods" and "Amortise To" are not yet used.
13. If you have used the same inputs as shown below – but with the £10,000 periodic payment, the G20 figure should currently equal £22,295.10.

	B	C	D	E	F	G
2	Amortisation Schedule					
3						
4	Amount		£90,000.00			
5	Number of Periods		8			
6	Interest Rate		2.50%			
7	Amortise to		£10,000.00			
8	Periodic Payment		£11,407.39			
9						
10						
11	Period	Opening Balance	Payment	Interest	Capital	End Balance
12						
13	1	£90,000.00	£11,407.39	£2,250.00	£9,157.39	£80,842.61
14	2	£80,842.61	£11,407.39	£2,021.07	£9,386.32	£71,456.29
15	3	£71,456.29	£11,407.39	£1,786.41	£9,620.98	£61,835.31
16	4	£61,835.31	£11,407.39	£1,545.88	£9,861.50	£51,973.80
17	5	£51,973.80	£11,407.39	£1,299.35	£10,108.04	£41,865.76
18	6	£41,865.76	£11,407.39	£1,046.64	£10,360.74	£31,505.02
19	7	£31,505.02	£11,407.39	£787.63	£10,619.76	£20,885.26
20	8	£20,885.26	£11,407.39	£522.13	£10,885.26	£10,000.00

14. To finish the calculation you need to make G20 equal the amount you want to amortise to. This can be done through trial and error, but using Goal Seek can help do this quickly.
15. Select Tools/Goal Seek. Fill in the required inputs. "Set Cell" should refer to G20, "To Value" should be the figure you want to amortise to and "By Changing Cell" refers to D8 – the periodic payment.

Goal Seek ? ×

Set cell:

To value:

By changing cell:

OK   Cancel

16. Choosing G20, 10,000 and D8 should give the result as shown above – a payment of £11,407.39.

That probably all seems pretty long-winded. The purpose of doing it, however, is to now show you how the PMT formula gives the same result – but much more quickly and offering more flexibility.

17. Insert the word "Payment" into F4 and insert the PMT formula into G4:

   G4=PMT(D6,D5,D4,-D7,0)*–1
   PMT(Rate, Number of Payments, Present Value, Future Value, Type)
   Rate = D6
   Number of Payments = D5
   Present Value = D4
   Future Value = -D7
   Type = 0 or can be omitted
   Multiply the formula by –1

18. This should result in the same figure as your Goal Seek found.

So, now you can confidently use the PMT formula to calculate your periodic outgoings – but is that enough? You will often need to know how much capital is outstanding at any given point in a cashflow. Of course, you could set up an amortisation schedule as shown above – but there is another way.

IPMT and PPMT

These two formulae work in a very similar fashion to the PMT formula – but split out the amount of interest (IPMT) and principal (PPMT) repaid in any given period.
   They can be used to replace the calculations in the E and F columns of the above example.
   The formulae take the following form:

   =IPMT(Rate, Period, Number of Payments, Present Value, Future Value, Type)
   =PPMT(Rate, Period, Number of Payments, Present Value, Future Value, Type)

And the inputs work in exactly the same way that the PMT inputs do. The only change is the second input – Period – as you need to tell the formula what period you are currently in.
   So, replace the calculations in E13 and F13 with:

   =IPMT($D$6,$B13,$D$5,$D$4,-$D$7)*–1
   =PPMT($D$6,$B13,$D$5,$D$4,-$D$7)*–1

Copy them down to row 20, and you will generate exactly the same payment schedule.
   Generally, when carrying out interest calculations, you will need to think carefully about whether you are treating payments as being at the start or end of each period. In the above examples, it is assumed that payments are made at the

end of each period – so during period 1, £2,250 of interest is accrued and paid off, with capital, at the end of the period.

This now gives you the flexibility to calculate the amount of interest and capital repaid in any given period – you no longer need to generate the whole table, just change the period input from being, in the above example, B13 (which is the current period for the row) to being a separate input option.

Capital outstanding

This leaves just one final element to take out of the schedule. So far, while we can calculate the amount of interest and capital repaid in any period, the end balance, being the amount of the loan outstanding, still needs details of the amount of capital repaid in each preceding period in order to be worked out.

We can get around this by using CUMPRINC, which is short for Cumulaitive Principal. This works as follows:

=CUMPRINC(Rate, Number of Payments, Present Value, Start Period, End Period, Type)

The inputs for which are largely the same as in previous formulae, save:

- **Present value** is the amount of capital to be repaid. As there is no Future Value input then this must equal the amount borrowed less the amount you wish to amortise to.
- **Start and end period** are the points between which you want to calculate the amount of capital repaid. These can be the same if you want the amount in a single period.
- **Type** is, as usual, used for timing of the payment (0 for payment at the end of the period, 1 for at the start). However, in this function, it must be included (in previous functions it could be omitted).

So, to utilise this in the above example:

19. Add a title, in column H, "Capital Repaid".
20. In H13, insert the CUMPRINC formula:

    H13: =CUMPRINC($D$6,$D$5,$D$4-$D$7,$B$13,B13,0)*–1

21. Note that the first B13 is fixed while the second is not. When copied down, the formula will always calculate the amount repaid from the first period to the current period (or you could just insert "1" as the first period!).
22. Copy the formula down to row 20 – the number in the 20th row should be £80,000 if you are still using the figures in the example – the net amount of capital to be repaid.
23. Now that we know the cumulative amount repaid in each period, we can work out the amount outstanding.
24. Add another column title in I – "Capital Outstanding".
25. All we need to do is deduct the value of H13 from the original balance:

    I13=$D$4-H13

26. Again, copy this down to row 20 and you should have a set of figures replicating the End Balance column. The extract below illustrates:

	E Interest	F Capital	G End Balance	H Capital Repaid	I Capital Outstanding
11	Interest	Capital	End Balance	Capital Repaid	Capital Outstanding
12					
13	£2,250.00	£9,157.39	£80,842.61	£9,157.39	£80,842.61
14	£2,021.07	£9,386.32	£71,456.29	£18,543.71	£71,456.29
15	£1,786.41	£9,620.98	£61,835.31	£28,164.69	£61,835.31
16	£1,545.88	£9,861.50	£51,973.80	£38,026.20	£51,973.80
17	£1,299.35	£10,108.04	£41,865.76	£48,134.24	£41,865.76
18	£1,046.64	£10,360.74	£31,505.02	£58,494.98	£31,505.02
19	£787.63	£10,619.76	£20,885.26	£69,114.74	£20,885.26
20	£522.13	£10,885.26	£10,000.00	£80,000.00	£10,000.00

Again, this is a long way to go simply to replicate numbers you already had. However, by now you should have an understanding of how these functions work in order that you can use them independently to calculate debt balances and other useful figures with confidence.

Chapter 10

Pulling it all Together

Creating an investment analysis cashflow

So far, we have looked separately at how to create an income cashflow, make allowances for costs and how to handle debt. In order to really make use of this, we have to pull all of the above into a single appraisal.

From this, we can generate a net income cashflow that we can analyse with *Net Present Value* and *Internal Rate of Return* calculations – more on these in the next chapter.

Please do not read this chapter with the idea that it is going to tell you how to set out a perfect cashflow, there are so many different cashflow applications that it is fruitless to try and set out a "textbook" example. In any event, you can already do this if you have worked through the previous chapters.

Rather, try and read this chapter as giving you a broad idea of how you might lay out a cashflow and the broad structure a cashflow can take.

You might like to dwell somewhat on the ideas for preparing the finance elements of the cashflow as these build on the previous chapter and will give you an example of how to incorporate finance without using the amortisation schedule.

Stage 1 – The inputs

It is much easier to write a spreadsheet if you have a clear idea of exactly what you want it to do and the degree of flexibility you will require. This will save you from having to try and add in additional features as an afterthought – this can be done, but is often quite difficult to do.

Consequently, giving some detailed thought to the inputs you require is time well spent.

For this example, we will assume that we are buying a multi-let building that we will finance, hold for a period and then sell. We might use the following overall assumptions – there will be more in the tenancy schedule:

	B	C	D	E	F	G	H	I	J	K	L
2	**Investment Purchase Example**										
3											
4	**Assumptions**										
5											
6	**Purchase**			**Professional Costs**			**Loan**			**Exit**	
7											
8	Purchase Price		£7,000,000	Rent Review Fee	10%		Loan to Value		60%	Sale Period	20
9	Initial Yield		7.84%				Loan		£4,200,000	Exit Yield	8.50%
10	Purchase Costs		5.75%				Equity		£2,800,000	Exit Costs	1.50%
11							Interest Rate		7.50%		
12	Rental Growth		3%				Loan Period		25 years		
13							Amortise to		£0		
14							Payment		£91,657		

Already there are some calculations to incorporate – some for later use, some just for information. The basic ones are as follows:

- D9 contains a calculation of the net initial yield – being the initial rent (you will see that this is in D24 below) divided by the purchase price and adjusted for purchasers costs:

D9=D24/D8/(1+D10)
Current Rent /Purchase Price/1+Purchase Costs

- J9 calculates the amount of borrowing:

J9=J8*D8
*Loan to Value * Purchase Price*

- J10 shows the amount of equity needed – simply

J10=D8-J9
Purchase Price – Loan

Once we get to J14, where the finance calculations start, is where we start to see the value of the finance formulae covered earlier. Rather than setting up a loan amortisation schedule, we calculate the necessary loan repayment as follows:

=PMT((1+J11)^0.25-1,J12*4,J9,-J13)*–1
*Periodic Payment (Quarterly Interest Rate, Number of Periods (Loan Period in years *4), Loan Amount, Amortise to J13)*–1 (just to make the figure positive)*

Stage 2 – The income cashflow

Below the inputs is a good place to set up your tenancy schedule – this can be as simple or complex as the task requires.

The following example includes many of the features discussed earlier.

	B	C	D	E	F	G	H	I	J	K	L	M	N	O	P
2	Investment Purchase Example														
3															
4	Assumptions														
5															
6	Purchase			Professional Costs			Loan			Exit					
8	Purchase Price		£7,000,000	Fees	1.7%		Loan to Value	60%		Sale Period	20				
9	Initial Yield	7.67%					Loan	£1,900,000		Exit Yield	8.50%				
10	Purchase Costs	5.75%					Equity	£3,900,000		Exit Costs	1.50%				
11							Interest Rate	7.00%							
12	Rental Growth	3%					Loan Period	20 years							
13							Amortise to	£0							
14							Payment	£91,652							
16	Income Cashflow													Annual Growth Index	
17													Operate	Void	Restart
18	Tenant	Area	Rent	psf	ERV	psf	Lease Details Review 1	Review 2	Review 3	Review 4	Break	Expiry	Break?	Period (Qtrs)	Date
20	First	30,000	£300,000	£10.00	£360,000	£12.00	24-Jun-04	24-Jun-09	24-Jun-14			24-Jun-14	24-Jun-17	Y	5	25-Dec-15
21	Second	20,000	£240,000	£12.00	£300,000	£15.00	25-Mar-05	25-Mar-10					25-Mar-15		5	24-Sep-15
22	Third	5,000	£40,000	£8.00	£45,000	£9.00	29 Sep 03	29 Sep 08					29 Sep 13		6	25 Mar 15
23															
24	Total	55,000	£580,000		£705,000										

The decision as to how complex your tenancy schedule needs to be will, naturally, vary with each application. Generally, it is best to add a little more than you expect to need (you will always be asked to allow for more scenarios than was originally envisaged) but do not go too wild. The more complex things get, the more time consuming to write and amend. Also, the more likelihood there is of errors creeping in.

Having set up the tenancy schedule, insert the quarter days and a period reference. The formulae further on will refer to periods that start at 1 in Q17 and dates that start in Q18.

Then take a deep breath and write the rental income cashflow – you should find enough information in previous chapters to do this.

	B	C	D	E	F	G	H	M	N	O	P	Q	R	DQ
15	Income Cashflow									Rental Growth Index		100.00	100.74	215.66
16														
17							Lease Details		Operate	Void	Restart	1	2	105
18	Tenant	Area	Rent	psf	ERV	psf	Review 1	Expiry	Break?	Period (Qtrs)	Date	25-Mar-05	24-Jun-05	25-Mar-29
19														
20	First	30,000	£300,000	£10.00	£360,000	£12.00	24-Jun-04	24-Jun-17	Y	2	25-Dec-14	75,000	75,000	128,316
21	Second	20,000	£240,000	£12.00	£300,000	£15.00	25-Mar-05	25-Mar-15		3	25-Dec-15	60,000	60,000	110,140
22	Third	5,000	£40,000	£8.00	£45,000	£9.00	29-Sep-08	29-Sep-13		4	29-Sep-14	10,000	10,000	15,922
23														
24	Total	55,000	£580,000		£705,000							145,000	145,000	254,380

Run the cashflow across for 105 periods or so (to column DQ) if you want to be able to see the loan, which is currently amortised over 25 years, decrease to zero.

If you have any trouble with the income cashflow, but still want to put together a cashflow framework, remember that you can always insert fixed figures for the income and worry about working through the income cashflow later. Do put in some rental increases at some point as you will need these changes in the next stage.

Stage 3 – Add the costs

Next, work out what costs you need to deduct – there may be none of any real significance if you are dealing with a long FRI lease but you may want to give some thought to potential rent review costs and the like.

In the assumptions, an allowance has been made for rent review fees at 10% of the new rent (renewals are being ignored as the assumed hold period is far shorter that the lease lengths). As discussed previously, you can complicate this as much as you need to. That is the advantage in dealing with costs in a separate cashflow. You can play around with them as much as you want without complicating your, already complex, income cashflow and the resulting figures are easy to see and check.

Set up your costs cashflow immediately below the income cashflow. In order to calculate the review costs (as the cashflow will end before the lease expiry dates) we can simply look for changes in the rental income flow to highlight reviews and calculate a charge from this.

Skipping the first quarter, which as mentioned previously is the easiest way to avoid an erroneous charge appearing at the start of the cashflow, R29 contains:

R29=IF(R20>Q20,R20*4*$G$8,0)*–1
IF the Current Quarter's rent is greater than Previous Quarter's rent then insert the Current Quarter's rent multiplied by four multiplied by the percentage fee, otherwise insert 0. Multiply value by –1 to make it negative.

You need to decide early on whether you want costs to be positive or negative. That sounds odd, but there is no reason that costs can not all be treated as positive figures in the cashflow provided that you remember to deduct them from, rather than add them to, the gross income line. Just remember to be consistent.

B	C	D	E	F	G	H	M	N	O	P	Q	R
15 Income Cashflow									Rental Growth Index		100.00	100.74
16												
17						Lease Details		Operate	Void	Restart	1	2
18 Tenant	Area	Rent	psf	ERV	psf	Review 1	Expiry	Break?	Period (Qtrs)	Date	25-Mar-03	24-Jun-03
19												
20 First	30,000	£300,000	£10.00	£360,000	£12.00	24-Jun-04	24-Jun-17	Y	2	25-Dec-14	75,000	75,000
21 Second	20,000	£240,000	£12.00	£300,000	£16.00	25-Mar-06	26-Mar-15		3	25-Dec-15	60,000	60,000
22 Third	5,000	£40,000	£8.00	£45,000	£9.00	29-Sep-03	29-Sep-13		4	29-Sep-14	10,000	10,000
23												
24 Total	55,000	£580,000		£705,000							145,000	145,000
25												
26												
27									Professional Costs Cashflow				
28												
29									First			0	0
30									Second			0	0
31									Third			0	0
32									Total			0	0

Stage 4 – Incorporate finance

We have assumed in the inputs above that this purchase will use debt to finance the purchase. That means we need to work out how much the cost of this will be on a quarterly basis and how much debt is outstanding at the end of the cashflow.

The first of these has already been dealt with in J14 and can now be incorporated. The only trick to remember is that the payment is, obviously, only made for a maximum of the number of payment periods so we need to start and stop the figure at the right time.

Assuming payments are in arrears, the payment will start in period 2, so R37:

R37=IF(R$17>$J$12*4+1,0,$J$14)*–1
IF current period is greater than the loan period in years times 4 plus 1, then insert 0, otherwise insert loan repayment figure.
Multiply by –1 to make cost negative.

Next, we will calculate the amount of the loan outstanding. As ever, there are several ways to do this. If you want to set up a complete amortisation schedule on a different page of the spreadsheet then do so. If you feel confident in using the financial formulae in isolation then use those. Here, we will use a blend of both methods.

First, work out how much of your payment in each period goes towards repaying the loan (as opposed to the interest component). In this case, R38:

R38=IF(R17>$J$12*4+1,0,PPMT((1+$J$11)^0.25–1,R17–1,$J$12*4,$J$9,-$J$13,0))*–1
IF current period is greater than the loan period in years times 4 plus 1, then insert 0, otherwise insert PPMT figure (see below).
Multiply by –1 to make it positive (if you want to).
PPMT(Rate, Period, Number of Payments, Present Value, Future Value, Type)
Rate = Rate in J11 converted to a quarterly figure
Period = Current Period –1 (as we are starting repayments in period 2)
Number of Payments = Loan period in years times 4
Present Value = Loan amount
Future Value = – amortise to amount
Type = 0 (in arrears)

This can then be copied across the cashflow. In row 39, you can keep a running total of how much of the loan is yet to be repaid:

R39=$J$9-SUM($R$38:R38)
Loan amount less the sum of the capital repaid in the first period to the current period

Note how only the first cell reference in the SUM formula is fixed. As this is copied across, it will add up everything from the first to the current quarter and deduct this from the loan amount.

	N	O	P	Q	R	S	T	U
15		Rental Growth Index		100.00	100.74	101.49	102.24	103.00
16								
17	Operate	Void	Restart	1	2	3	4	5
18	Break?	Period (Qtrs)	Date	25-Mar-03	24-Jun-03	29-Sep-03	25-Dec-03	25-Mar-04
19								
35	**Finance Cashflow**							
36								
37	Loan Repayment				-91,657	-91,657	-91,657	-91,657
38	Capital Repaid				15,030	15,304	15,583	15,868
39	Capital Outstanding				4,184,970	4,169,666	4,154,083	4,138,215
40								
41	Net Income			145,000	53,343	50,244	54,845	54,845

Why use this method, rather than just the financial formulae? Well, if you copy this cashflow out over the entire extent of the loan, you will be able to see that in the 101st period (i.e. after 100 payment periods) the loan outstanding has reduced to zero. This should give you confidence that the calculation is working properly.

If you want to avoid using the cashflow, you could include the loan outstanding calculation in the input area – just as with the calculation of the quarterly loan repayment.

	H	I	J	K	L
6	**Loan**			**Exit**	
7					
8	Loan to Value		60%	Sale Period	20
9	Loan		£4,200,000	Exit Yield	8.50%
10	Equity		£2,800,000	Exit Costs	1.50%
11	Interest Rate		7.50%		
12	Loan Period		25 years	Loan Outstanding	£3,862,320
13	Amortise to		£0		
14	Payment		£91,657		

In the above, L12:

L12=J9+CUMPRINC((1+J11)^0.25–1,J12*4,J9,1,$L$8–1,0)
Loan Amount + Principal repaid at the sale date (added together as this is a negative figure)
CUMPRINC(Rate, Number of Payments, Present Value, Start Period, End Period, Type)
Rate = Rate in J11 converted to a quarterly figure
Number of Payments = Loan in years times 4
Present Value = Loan Amount
Start Period = 1
End Period = Sale Period –1 (as if we hold the investment for 20 periods and pay in arrears at the end we have only made 19 payments)
Type = 0 (payments in arrears)

Having done this, you can then total your net income in row 41.

Stage 5 – Incorporate capital movements

Having finished the net income calculation, you may need to incorporate sale and purchase prices – depending on the use to which you are putting the cashflow.

The purchase figures will usually be fairly straightforward and will appear as a one-off cost in the first period:

Q46=D8*–1
*Purchase Price * –1*
Q47=Q46*D10
*Purchase Price * Purchase Costs*
Q48=J9
Q48 = Loan Amount

Which results in the following (note the hidden columns):

	N	O	P	Q	AJ
17	Operate	Void	Restart	1	20
18	Break?	Period (Qtrs)	Date	25-Mar-03	25-Dec-07
19					
44	**Capital Movements**				
45					
46	Purchase Price			-£7,000,000	
47	Purchase Costs			-£402,500	
48	Loan			£4,200,000	
49					
50	Sale Proceeds				£8,265,446
51	Sale Costs				-£123,982
52	Repay Loan				-£3,862,320
53					
54	**Overall Cashflow**			-£3,057,500	£4,187,487

You will need to either reduce the purchase price by the amount of the loan, or add back in the loan amount, as shown above, to ensure that the net starting payment is correct.

You may also need to consider what additional purchase costs must be deducted, such as those relating to bank fees and so forth, but for the purposes of this example we will not worry about these.

When it comes to the exit price you will need to consider how best to value the property at the future date. The simplest approach is to apply a net initial yield to the current income at that point in time.

In the above example:

AJ50=IF(AJ$17=$L$8,AJ$24*4/$L$9/1.0575, "")
IF current period = Sale Period, Net Income*4/Exit Yield/1.0575, otherwise insert a blank
The 1.0575 division adjusts the value for purchaser's costs and will likely be market driven but this could refer to an input cell if you need to vary these.
AJ51=IF(AJ50<> "",AJ50*$L$10*–1, "")
*IF Sale Proceeds are anything other than a blank, insert Sale Proceeds * Exit Costs *–1, otherwise insert a blank*
AJ52=IF(AJ$17=$L$8,AJ$39*–1, "")
*If current period = Sale Period then insert Capital Outstanding *–1, otherwise insert a blank.*

Of course, the above were originally entered in column Q and copied along the length of the cash flow.

You may also want to calculate what the reversionary yield is at that point to help gauge whether the value is about right.

In the above example this can be done with straightforward multiplication using the rent index at the sale date and the total ERV figure.

There are numerous ways that the exit value can be calculated and the choice will depend on your circumstances and assumptions at the time.

Stage 6 – Total it up

Not difficult to do – but not quite a simple as it sounds. You need to bear in mind when payments are received. Usually, this will mean that you don't want to add in the rental income in the final quarter, as this will go to the new purchaser if rent is payable in advance.

So, in Q54, as shown above:

=IF(Q17>$L$8,"",IF(Q17=$L$8,SUM(Q52,Q51,Q50,Q48,Q47,Q46,Q37,Q32),SUM(Q52, Q51,Q50,Q48,Q47,Q46,Q41)))

IF current period is greater than Sale Period then insert a blank, IF the current period = the Sale Period then insert the sum of all costs and income excluding current quarter's rent, otherwise insert sum of all costs and income.

Why stop the figures after the sale period? This helps us, for example, when we come to work out internal rate of return figures later.

You now have a completed cashflow that incorporates income, costs and finance as well as the purchase costs and net sale proceeds.

Chapter 11

Utilising Cashflows

So now what are you going to do with it?

Once you have set up your income cashflow, what next? How are you going to interpret the results? The good news is that the hard part is out of the way, but so far all you have is a list of numbers – we need to make them work for us.

Descriptive cashflows

Security of income

Perhaps the simplest application is to use an income cashflow to illustrate how secure the income from an investment is.

When considering a portfolio or a property that is let to a large number of tenants, it is often difficult to gain a clear impression of how secure the income flow is as a balance has to be pictured between the unexpired term of each lease and the proportion of rent that is secured against that tenancy.

Earlier, we created a cashflow that generated income for just the period of the current tenancy, with an uplift to current ERV if appropriate:

	B	C	D	E	F	G	H	I
						25-Mar-03	24-Jun-03	29-Sep-03
2	Tenant	Rent	ERV	Review	Expiry			
3	Smith	5,000	5,500	24-Jun-03	24-Jun-08	1,250	1,375	1,375
4	Jones	6,000	6,500	24-Jun-04	24-Jun-09	1,500	1,500	1,500
5	Brown	7,000	6,000	24-Jun-03	24-Jun-08	1,750	1,750	1,750
6	Total					4,500	4,625	4,625
7	Annual Basis					18,000	18,500	18,500

If you simply add a total income line (multiplying by four if you want to annualise the income) and run this along the length of the cashflow, then graphing this shows:

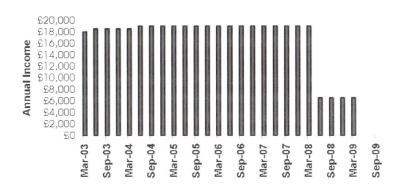

Which gives a good impression of how long the income from the property lasts.

Running yields

The security of income idea can be extended to demonstrate the running return offered by the investment.

If the gross purchase price of the property were £250,000, remembering that the gross price includes not only the purchase price of the investment but all associated costs, then the running yield can be found by simple division.

Here, G7=G6/$E$7 to give a percentage:

	B	C	D	E	F	G	H	I
2	**Tenant**	**Rent**	**ERV**	**Review**	**Expiry**	**25-Mar-03**	**24-Jun-03**	**29-Sep-03**
3	Smith	5,000	5,500	24-Jun-03	24-Jun-08	1,250	1,375	1,375
4	Jones	6,000	6,500	24-Jun-04	24-Jun-09	1,500	1,500	1,500
5	Brown	7,000	6,000	24-Jun-03	24-Jun-08	1,750	1,750	1,750
	Total					4,500	4,625	4,625
6	**Annual Basis**					**18,000**	**18,500**	**18,500**
7	**Gross Purchase Price**			£250,000		7.20%	7.40%	7.40%

Again, this can be graphed:

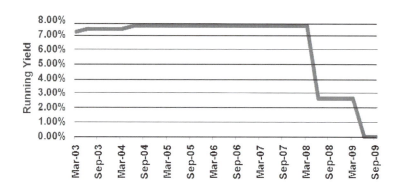

Appraisal cashflows

The above offers a couple of examples to allow you to use cashflows for illustrative purposes. Below, we look at how a cashflow can be used as an appraisal tool.

Net present values

A net present value (NPV) calculation takes the cashflow from an investment, applies a discount rate to convert future figures into current values and adds them all together.

NPVs can be used to create valuations from cashflows. The Excel spreadsheet has an in-built formula to calculate these – but, as with the financial functions, it is reassuring to have an idea of how this works.

Let's use a simple example – what is the value of £1,000 received at the end of each year for 5 years at a discount rate of 8%?

This can easily be calculated using a spreadsheet:

	B	C	D	E	F	G
2		1	2	3	4	5
3	Income	£1,000	£1,000	£1,000	£1,000	£1,000
4	Present Value	£925.93	£857.34	£793.83	£735.03	£680.58
5						
6	Discount Rate		8%			
7	Total		£3,992.71			

Having set out the cashflow in C3..G3, each future sum is discounted to present value by multiplying it $(1+i)^{-n}$ where i is the discount rate and n is the number of periods to discount over. So:

C4=C3*(1+$D$6)^–C2
Payment in period(1+Discount Rate)^–Number of Periods to discount*

Copying this across the cashflow, as far as G4, and totalling the results in D7 gives a total of £3,992.71.

However, the simplest method is to use the built in NPV function within Excel. This takes the format:

=NPV(rate,value1,value2, ...)
Rate is the Discount Rate on the same periodic basis as your cashflow (i.e. annual, quarterly etc)
Value can either be individual figures or a range

The NPV formula does all of the hard work for you. So, you can delete the Present Value line and insert the NPV formula into D7:

D7=NPV(D6,C3:G3)
NPV(Discount Rate, Cashflow Range)

Which generates the same result as the previous approach, a figure of £3,992.71.

	B	C	D	E	F	G
2		1	2	3	4	5
3	Income	£1,000	£1,000	£1,000	£1,000	£1,000
4						
5						
6	Discount Rate		8%			
7	NPV		£3,992.71			

One important point to note with the NPV calculation is that it treats all income as if it is received at the end of the period – i.e. payment in arrears. If you are receiving payments at the start of the period – such as rent – you would not want to discount the first payment. So, you need to adjust the formula so that the NPV range, in the above, only refers to years 2 to 5, to which you add the first income payment:

=NPV(D6,D3:G3)+C3

In the same fashion, you can apply an NPV to a far more complex cashflow – such as the example discussed in the previous chapter.

In the example, the end result was a net cashflow running from Q54:AJ54 (if a sale period of 20 was used). To calculate an NPV of this – assuming quarterly payments at the beginning of each period, you could add a results section within the assumptions area of the cashflow – perhaps adjacent to the Exit assumptions:

N8=Discount Rate
O8=8%
N9=NPV
O9=NPV((1+O8)^0.25–1,R54:DQ54)+Q54
NPV((1+Annual Rate)^0.25–1, Net Cashflow excluding first Quarter)+First Quarter

	N	O
6	**Results**	
7		
8	**Discount Rate**	8.00%
9	**NPV**	£994,357

In the above, the NPV range goes as far as column DQ – which is the furthest the cashflow goes to. By doing this you can be sure that as you change the sale date assumptions, your NPV calculation will include the whole cashflow. We have already made sure that the overall cashflow only includes figures up to and including the sale date.

You need to think about how to interpret the result. In the above example, as you have included the initial purchase price in the NPV calculation, the result will show what your net gain/loss will be on the investment based on your chosen discount rate.

In the next chapter we will use the NPV function to compare the value of different income streams.

Internal Rate of Return

The Internal Rate of Return (IRR) is the rate at which the total of the discounted values of all future payments are equal to the price paid for an investment. It is therefore closely linked with the NPV calculation.

As the previous NPV example includes the purchase price of the investment, you could calculate the investments IRR by trial and error – by adjusting the discount rate until the NPV equals zero.

However, Excel will do this for you – using, as you might expect, the IRR function, which works as follows:

=IRR(Value Range, Guess)
Value range covers the cashflow that you wish to calculate the IRR for and must contain at least one positive and one negative value.
Guess provides the IRR with a start figure from which to begin its iterative calculation. If left blank then 10% will be assumed.

By way of example we will calculate an IRR in two different ways. Consider an investment for which you pay £3,000, receive regular payments of £500 and then get your original stake back. We can set this up as follows:

	B	C	D	E	F	G
2		1	2	3	4	5
3	Income	-£3,000	£500	£500	£500	£3,000

To calculate the IRR you could use the NPV function and keep changing the discount rate until you get a zero answer. Prepare an input cell for your discount rate in D5, use 10% to start with, and insert the NPV calculation as follows:

D6=NPV(D5,D3:G3)+C3
NPV(Rate, Cashflow excluding first payment)+First Cashflow Payment

This gives:

	B	C	D	E	F	G
2		1	2	3	4	5
3	Income	-£3,000	£500	£500	£500	£3,000
4						
5	Discount Rate		10.00%			
6	NPV		£292.47			

By adjusting the discount rate until the NPV equals zero you can calculate the IRR – which in this case is around 13.24%.
To get Excel to do the trial and error for you, insert:

D8=IRR(C3:G3,0.1)
IRR(Cashflow block, IRR guess at 10%)

This will confirm your answer.

	B	C	D	E	F	G
2		1	2	3	4	5
3	Income	-£3,000	£500	£500	£500	£3,000
4						
5	Discount Rate		13.24%			
6	NPV		£0.00			
7						
8	IRR		13.24%			

Once again, it is no more difficult to apply this to a more complex cashflow. Considering the example in the previous chapter the IRR of the cashflow could be calculated in the results section by using:

N11=IRR
O11=(1+IRR(Q54:DQ54,0.1))^4−1
(1+(IRR(Entire cashflow block, 10% guess))^4−1

The only thing that slightly complicates the formula is that, as the cashflow is on a quarterly basis, the result needs to be converted to an annual rate and the above does this.

	N	O
6	Results	
7		
8	Discount Rate	8.00%
9	NPV	994,357
10		
11	IRR	15.73%

Sensitivity

When calculating an NPV, what discount rate should you use? This will depend on the use to which you are putting the cashflow and your assumptions. However, often you will want to examine the effects of changing this rate.

You can do this using the Excel "Table" function which allows the impact of changing variables to be examined without having to manually amend a calculation. Below is a simple example, but it is no more difficult to apply this to a complex cashflow – once you know where the inputs go.

Set yourself up some figures to carry out your sensitivity analysis on (see over). Here, we will use a basic valuation calculation so input some labels, as shown below, into B2 to B4, rent and yield figures in column C and a formula in C4:

C4=C2/C3

Below this we will create the sensitivity analysis that examines the effect of changing two variables. As you can see, the variations are to the rent and yield figures, so set up the changing rent in row 6 and the varying yield in column B.

When we use a data table, we need to give the spreadsheet a result to show. The formula for this result goes at the juncture of the two variables. Here, B6 contains:

B6=C4

So far, you should have:

	B	C	D	E	F	G
2	Rent	£1,000				
3	Yield	8%				
4	Value	£12,500				
5						
6	£12,500	£900	£950	£1,000	£1,050	£1,100
7	7.00%					
8	7.50%					
9	8.00%					
10	8.50%					
11	9.00%					

To carry out the analysis, first select the entire table – in this case click in B6 and drag to highlight out to G11. Then, from the Data menu, choose Table:

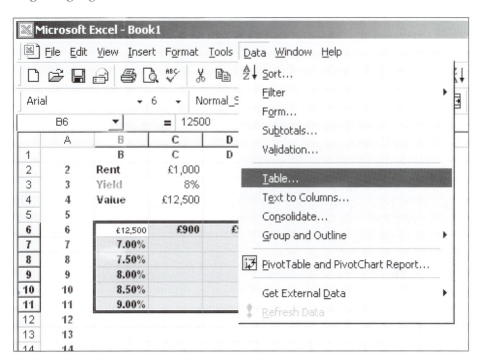

This will bring up the following requester:

After the figure, the Table requester shows:

Table

Row input cell:

Column input cell:

OK   Cancel

This first asks for the row input cell, or in other words "Which cell in your spreadsheet should be varied with the numbers from the top data table row?" – in this case it is C2. Similarly, the column input cell is C3.

Having selected "OK" the following result will be generated (which has been formatted to tidy up the presentation):

	B	C	D	E	F	G
2	Rent	£1,000				
3	Yield	8%				
4	Value	£12,500				
5						
6	£12,500	£900	£950	£1,000	£1,050	£1,100
7	7.00%	£12,857	£13,571	£14,286	£15,000	£15,714
8	7.50%	£12,000	£12,667	£13,333	£14,000	£14,667
9	8.00%	£11,250	£11,875	£12,500	£13,125	£13,750
10	8.50%	£10,588	£11,176	£11,765	£12,353	£12,941
11	9.00%	£10,000	£10,556	£11,111	£11,667	£12,222
12						

Having created the table, changes made to the row and column inputs will result in the table being automatically updated – try changing the £900 start figure and see what happens. For this reason it is worth linking the sensitivity input figures using a formula – for example:

D6=C6+50

Once copied across, varying the initial analysis input will update the entire table.

Once you can perform a simple sensitivity analysis you can apply this to anything. However complex your spreadsheet gets this technique does not really change. The only complexity tends to be the requirement to link other inputs together.

In the above, with only one rent and yield, it is easy to apply the variations. If you had a multi-let property which has a range of ERVs and you wanted to vary the rent by increasing increments you might have to link all of your rental inputs together and then perform your sensitivity on the base ERV – or on the variance between ERVs.

For example, if you had three grades of office space and your ERV was £100 psm for worst space, £120 psm for average accommodation and £140 psm for the best. Rather than inputting these as separate numbers you could link them so that average equals worst plus £20 and best is worst plus £40. Then, as you vary the worst input the change will impact on the remainder of the calculation even though you are only changing one variable.

Alternatively, you could set up a variable increment – in this case £20 and link them like this – average equals worst plus increment, best equals average plus increment. You could then analyse what happens if the gap between the grades of space opens and closes by performing your sensitivity on the incremental figure rather than on the rent.

Chapter 12

Variable Rent Reviews

Incorporating reviews on different bases

Increasingly there is demand for more flexible leases. Not only are there requirements for shorter overall terms and the inclusion of break options, but also for rents to be able to fall on review if the market has weakened since the grant of the lease.

Consequently, there is the increasing need to be able to incorporate the potential for downwards rent reviews within cashflows.

This is simple enough to do, indeed in isolation it makes the income cashflow formulae simpler as it removes the need to check whether ERV is greater than passing rent. However, landlords are not always happy to agree to such terms and may seek to limit the scope of any downwards movement in rent – perhaps to no lower than the initial rent rising in line with inflation.

Also, it requires consideration to be given to pricing. What initial rent should be charged for a traditional 10-year lease with upwards-only reviews as opposed to the same lease with the potential for a fall in rent at the fifth year?

The following looks at an example of how to assemble a rental cashflow that can compare different rental streams due to varying rent review provisions. We will work on a single square foot of rent for the sake of simplicity.

Stage 1 – The inputs

The following illustrations show a suggested layout for this example, which can be easily copied (see over).

The first inputs are for rental growth. It is possible that there may be a difference between what is regarded as market value for a property and the initial rent. So an input is provided upon which to base the rental growth.

Space is then included for 10 years of rental growth forecasts. The assumption is that the final year's figure (2013 in this case) will be used as an average for future growth. The somewhat curious rates incorporated have been chosen purely to help illustrate how this technique works.

An input is then included for a "Base Increase". In this example, there will be an option to review either to pure market value or to a minimum of the indexed base rent – you might use the retail price, or similar, index.

The tenancy schedule is placed below this. Enter the headings as shown and then insert the quarter days (using the usual formula) starting in T16 and extending to DZ16. This will give room for a 25-year cashflow with a few extra dates to help with the lookup formula that will be used below.

You can enter the tenancy assumptions in row 20, as shown, by simply typing in the appropriate details. However, the example will work better if we automate parts of the schedule.

Enter the input figures as shown under the headings in columns B to F, K to N, P and R.

	B	C	D	E	F	G	H	I	J	K	L
2	**Variable Review Example**										
3											
4	Rental Growth Assumptions										
5											
6	Current ERV			£11.00							
7											
8	Year	2004	2005	2006	2007	2008	2009	2010	2011	2012	2013
9	Rate	5%	5%	5%	5%	5%	-10%	-10%	-10%	-5%	5%
10											
11	Base Increase		3.00%								
12											
13											
14											
15											
16											
17			Lease	Review	Lease		Rent Reviews			Up and Down	Base Index
18	Option	Rent psf	Term	Period	Start	First	Second	Third	Fourth	Review?	Minimum?
19											
20	One	£10.00	25	5	25/12/2003	25/12/2008	25/12/2013	25/12/2018	25/12/2023	y	n

	M	N	O	P	Q	R	S	T	U	V
2										
3										
4										
5										
6										
7										
8					Rental Growth Index			100.00	101.23	102.47
9					ERV			£11.00	£11.13	£11.27
10										
11					Base Index			100.00	100.74	101.49
12					Base Rent			£10.00	£10.07	£10.15
13										
14					Annualised Rent			£10.00	£10.00	£10.00
15										
16								25-Dec-03	25-Mar-04	24-Jun-04
17	Break	Operate	Lease Expiry	Void	Reletting	Discount	NPV of			
18	Date	Break?		Period (Qtrs)	Date	Rate	Income			
19										
20	25/12/2012	n	25/12/2028	3	29/09/2029	10%		2.50	2.50	2.50

Next calculate the expiry date of the lease in O20. The reason for calculating this before the review dates is because the result of this cell will be used in the rent review date formulae:

O20=IF($N20="y",$M20,DATE(YEAR($F20)+$D20,MONTH($F20),DAY($F20)))

IF Operate Break=y, insert Break Date, otherwise insert DATE(Lease Start Year plus Lease Term, Lease Start Month, Lease Start Day)

Now for the review dates:

G20=IF(F20="","",IF(YEAR($O20)>YEAR(F20)+$E20,DATE(YEAR(F20)+$E20, MONTH(F20),DAY(F20)), ""))

IF previous cell is blank, then insert a blank, IF the Lease Expiry year is greater than the previous cell year plus the Review Period, then insert DATE(previous cell year plus review period, previous cell month, previous cell day), otherwise insert a blank.

Copying this across to columns H, I and J will calculate up to four review dates. As you adjust the lease length you will be able to see these appear and disappear as required.

Finally, calculate the reletting date:

Q20=HLOOKUP(O20+(P20*98),$T$16:$DZ$16,1)
HLOOKUP the Lease Expiry date plus the Void Period in quarters times 98 in the quarter day block and return a result from that row.

As usual, the 98 in the above formula is a number slightly bigger then the number of days in a quarter to ensure that the HLOOKUP formula goes slightly beyond the date we require.

We will return to the NPV calculation in S20 later.

Stage 2 – The indices

You can see that there are headings for the indices in column Q – rows 8–14. Fill these in. Also, add the base figures for the indices – 100 goes in cells T8 and T11.

First, we will calculate the rent index, using the LOOKUP technique:

U8=(1+LOOKUP(YEAR(U16),$C$8:$L$8,$C$9:$L$9))^0.25*T8
LOOKUP the year of current quarter in the rental growth year block, return the corresponding value from the rental growth rate block. Add 1 to this. Raise resulting figure to the power of 0.25 and multiply by the previous index figure.

Copy this across to column DZ.
Now we can work out the ERV in each quarter:

T9=T8/100*$E$6
*Current Rental Growth Index/100*Current ERV input*

Again this is copied across to column DZ.

The Base Index is simpler to calculate as we are only using a single input – although there is no reason why you could not incorporate the same flexibility as the rental growth index. In U11:

U11=T11*(1+$D$11)^0.25
Previous Index figure(1+Base Increase)^0.25*

A calculation showing what the initial rent inflated in line with the Base Index will be in each quarter can now be inserted:

T12=T11/100*$C$20
*Base Index/100*Initial Rent*

The line for Annualised Rent is for graphing purposes and will be used later.

Stage 3 – Rental cashflow

The income cashflow formula takes largely the same form as previously used, starting at the end and working backwards, although the review uses the MAX function, rather than an IF test so we will look at this first.

The tenancy schedule contains two "switches" relating to the rent review. IF statements can be used to select which elements are included within the review and MAX can be used to select the highest.

=MAX(IF($L20="y",U$12/4,0),U$9/4,IF($K20="n",T20,0))
Choose the maximum from:
IF Base Index Minimum="y" include current Base Rent/4, otherwise 0
Current ERV/4
IF Up and Down Review="n" include previous quarter's rent, otherwise 0

This is incorporated within the following formula:

U20=IF(U$16=$Q20,U$9/4,IF(U$16=$O20,0,IF(OR(U$16=$G20,U$16=$H20,U$16=$I20, U$16=$J20),MAX(IF($L20="y",U$12/4,0),U$9/4,IF($K20="n",T20,0)),T20)))
IF Current Quarter = Reletting Date then insert ERV/4, IF Current Quarter = Lease Expiry then insert 0, IF Current Quarter = any of the Review Dates, then insert the results of the above MAX formula, otherwise copy the previous quarter's rent.

Copy this across the cashflow and then finish by inputting the rent for the first quarter of the cashflow:

T20=C20/4
Initial Rent/4

Once that is completed fill in the annualised rent row:

T14=T20*4
*Current quarter's cashflow * 4*

Stage 4 – Graph the results

In order to make sure the cashflow works properly set up a graph that shows the ERV, Base Rent and Annualised Rent – rows 9, 12 and 14.
This will generate the following result:

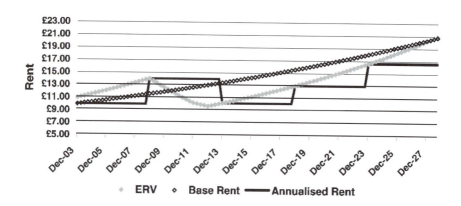

It is worth changing the review assumption switches to see the impact that this has. Changing the Base Rent Minimum option to "y" generates the following result:

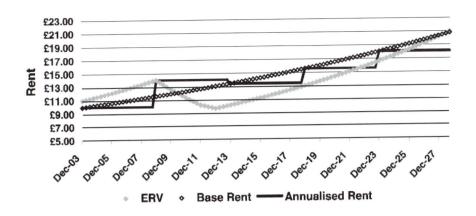

Stage 5 – Comparing results

By introducing an NPV calculation into the spreadsheet you can use this cashflow to compare different letting scenarios.

First we will set up the NPV calculation:

S20=NPV((1+$R20)^0.25-1,U20:DO20)+T20

NPV(Discount Rate converted to a quarterly amount, cashflow block excluding first quarter)+first cashflow quarter.

Next, copy the whole row 20 down one row so that you have two scenarios.

As you amend the inputs, you can compare the relative values of the two cashflows using the NPV figure.

Chapter 13

Development Appraisals

Some techniques to help with common problems

Development appraisals generally accumulate costs and calculate the interest charge that is accrued over the life of the development in order to enable the overall profit that is expected from a scheme to be calculated.

This is a simplified view of such appraisals and there are a myriad of complications that can be added. For example, profit shares and overage calculations may need to be incorporated to model an individual transaction and this can complicate matters.

However, even within a simple development appraisal there are several areas where flexibility is sometimes difficult to incorporate. Set out below are some thoughts on these.

General costs

Development cashflows are often set up on a monthly or quarterly basis – usually depending on the length of the project. In the following, we will use the former. Also, as well as labelling the cashflow with dates, it is often helpful to include a period reference, as shown in the extract below.

Dealing with one off costs can be done using an IF statement. For example, if you need to incorporate an architect's fee that is based on a proportion of the build cost and is all to be paid in the second period, a single IF statement would do the job. In the example below:

D11=IF(D8=$E$5,$D$5,0)
IF Current Period = Charging Period then insert cost, otherwise insert 0.

	B	C	D	E	F	G	H	I	J
2	Build Cost	£1,000,000							
3									
4		Charge	Amount	Period					
5	Architect	5%	£50,000	2					
6									
7									
8		Period	1	2	3	4	5	6	7
9		Date	Jan-04	Feb-04	Mar-04	Apr-04	May-04	Jun-04	Jul-04
10									
11	Architect		£0	£50,000	£0	£0	£0	£0	£0

Naturally, you will want to incorporate a calculation in D5 to allow you to vary the amount charged if the build cost or fee rate changes:

D5=C5*C2
*Fee Rate * Build Cost*

However, if costs are to be on a staged basis then the formulae can become quite complex. If the fee were to be payable in tranches, you might set up your inputs as follows:

	B	C	D	E	F	G	H	I	J
2	Build Cost	£1,000,000							
3									
4		Charge	Amount	2	6				
5	Architect	5%	£50,000	25%	75%				
6									
7									
8		Period	1	2	3	4	5	6	7
9		Date	Jan-04	Feb-04	Mar-04	Apr-04	May-04	Jun-04	Jul-04
10									
11	Architect		£0	£12,500	£0	£0	£0	£37,500	£0

E4 and F4 contain the period reference when tranches are to be paid out, E5 and F5 contain the proportion of the fee. If, as in this case, there are only two payments then IF statements might still be appropriate:

D11=IF(D8=$E$4,$E$5*$D$5,IF(D8=$F$4,$F$5*$D$5,0))
*IF Current Period = First Tranche Period, insert First Tranche Rate * Amount, IF Current Period = Second Tranche Period then insert Second Tranche Rate * Amount, otherwise insert 0.*

However, if the payment is broken down into more stages then this will get cumbersome quite quickly. An alternative approach – which looks complicated for only two payments but gets no more complex as additional tranches are added in – is to use HLOOKUP.

You can use HLOOKUP to extract the percentage to be paid in the appropriate quarter as follows:

D11=HLOOKUP(D8,$E$4:$F$5,2,FALSE)
Look for the Current Period in the first row of the payment period block and return the result from the row below. Only return a result if an exact match is found.

If you enter this and copy it across you will see the following:

	B	C	D	E	F	G	H	I	J
2	Build Cost	£1,000,000							
3									
4		Charge	Amount	2	6				
5	Architect	5%	£50,000	25%	75%				
6									
7									
8		Period	1	2	3	4	5	6	7
9		Date	Jan-04	Feb-04	Mar-04	Apr-04	May-04	Jun-04	Jul-04
10									
11	Architect		#N/A	25%	#N/A	#N/A	#N/A	75%	#N/A

While the formula correctly inserts the percentages into the cashflow, if an exact match cannot be found then an error is generated. To avoid this, we can use a single IF statement and a new function called ISNA.

ISNA allows you to query whether a formula gives a "#N/A" error (number not available). If it does, ISNA returns a TRUE result, if it does not then a FALSE result is given.

The format for this is simply:

=ISNA(calculation or cell reference)

We can use this within an IF statement as follows:

D11=IF(ISNA(HLOOKUP(D8,$E$4:$F$5,2,FALSE)),0,HLOOKUP(D8,$E$4:$F$5,2,FAL SE))*$D$5

IF ISNA(HLOOKUP function) gives a TRUE result then insert 0, otherwise insert the result of the HLOOKUP function. Multiply by the Fee Amount.

Note that as the ISNA function by definition equals TRUE or FALSE there is no need to put "=TRUE" (i.e. =IF(ISNA(HLOOKUP(D8,$E$4:$F$5,2, FALSE))=**TRUE**, ...") after the ISNA function. The IF statement will simply jump to the appropriate result without this.

Using this technique, if you had more payment tranches you would simply add to the input block and extend the range of the HLOOKUP function to cover these.

Spreading the build cost

Central to a development is the cost of construction. Aside from the total cost of construction there are three variables which, when combined, require a degree of ingenuity to incorporate.

These factors are:

- When the construction work is due to start.
- How long the period of construction will be.
- Costs apportioned on an S-Curve basis (which varies with the length of the construction period).

There are various ways to deal with this. One reasonably simple way, based around the HLOOKUP function, is described below.

S-Curve table

First you will need to set up a table of the S-Curves you are likely to need. This will depend on how much variation you expect to need in your build period – i.e. if you expect your build period to take nine months then it is not a bad idea to set up the data for six to 12 months.

This gives you a matrix from which to extract the appropriate percentage of build cost based on the length of construction and the current period.

N	O	P	Q	R	S	T	U	V	W	X	Y	Z
3 Period	1	2	3	4	5	6	7	8	9	10	11	12
4 6 months	9%	13%	28%	28%	13%	9%						
5 7 months	7%	10%	17%	32%	17%	10%	7%					
6 8 months	5%	7%	14%	24%	24%	14%	7%	5%				
7 9 months	4%	6%	10%	16%	28%	16%	10%	6%	4%			
8 10 months	3%	5%	9%	12%	21%	21%	12%	9%	5%	3%		
9 11 months	2%	3%	6%	11%	14%	28%	14%	11%	6%	3%	2%	
10 12 months	1%	2%	5%	9%	14%	19%	19%	14%	9%	5%	2%	1%

Extracting the build cost

Rather than set up the entire appraisal, we will just use some selected inputs to demonstrate how this will work:

	B	C	D	E	F	G	H	I
3	Build Cost	£100,000						
4	Build Start	2						
5	Build Period	6						
6								
7	1	2	3	4	5	6	7	8
8								
9								

Development will not always commence in period 1. So, the first trick is only to make the build cost appear in the appropriate range of cells.

To do this, we need a formula along the lines of "**IF** the **current period** lies between the **start of the build period** and the **end, then** put a cost, **otherwise** do nothing."

Based on the above layout, we can do this as follows (do try it yourself first):

B9=IF(AND(B$7>=$C$4,B$7<$C$4+$C$5), "Here", "")
IF the current period is greater than or equal to the Build Start AND the current period is less than the Build Start plus the Build Period then insert "Here", otherwise insert a blank.

Which will produce this result once copied across:

	B	C	D	E	F	G	H	I
3	Build Cost	£100,000						
4	Build Start	2						
5	Build Period	6						
6								
7	1	2	3	4	5	6	7	8
8								
9		Here	Here	Here	Here	Here	Here	

As you change the start and build periods then the "Here's" will move up and down the cashflow.

Inserting the build cost

Now we know where the build cost will go, we just have to look up the appropriate S-curve multiplier and apply it to the total build cost.

You will recall that HLOOKUP requires a value to search for, a block of data and the number of rows to look down the table. In this instance, these are as follows:

- **Value to find** – You need to search along the build period at the top of the S-curve data block to find the period that you are in. In this example it will be equal to the **Current Period – Build Start + 1**. So, if you are in period 5, and your build starts in period 5, then you will be searching for 1 in the table.
- **Data block** – This is simply the S-curve data, including the period numbers at the top – i.e. O3:Z10.
- **Rows down** – In the example, a six-month build is in row 2 of the data block (remembering that this is different to it being in row 4 of the spreadsheet) – so you would need **Build Period – 4**. Searching for periods outside of the data block will produce errors – hence the need for consideration of the likely range of variance early on.

So, putting that together, gives:

HLOOKUP(C$7-$C$4+1, $O$3:$Z$10, $C$5-4)*$C$3

Lookup how far you are into the build period in the S-curve block and return the percentage in the appropriate build period row. Multiply by the total build cost.

And integrating that into the positioning formula, replacing the "Here" comment, gives:

B9=IF(AND(B$7>=$C$4, B$7<$C$4+$C$5), HLOOKUP(B$7-$C$4+1, $O$3:$Z$10,$C$5-4)*$C$3,"")

Which will generate the following:

	B	C	D	E	F	G	H	I
3	Build Cost	£100,000						
4	Build Start	2						
5	Build Period	6						
6								
7	1	2	3	4	5	6	7	8
8								
9		£9,000	£13,000	£28,000	£28,000	£13,000	£9,000	

Interest calculation

Once you have entered all of your costs, you will be able to add up the column for each month and generate a total expenditure on a month-by-month basis. You will then need to incorporate the interest rate calculation.

Assuming that interest is charged in arrears you could do this as follows. Below the Total Expenditure calculation add two further lines – Interest in Period and Cumulative Cost.

	D	E	F	G	H	I	J	K
10	Interest Rate	8%						
11								
12	Period	1	2	3	4	5	6	7
13	Date	Jan-04	Feb-04	Mar-04	Apr-04	May-04	Jun-04	Jul-04
14								
15	Total Expenditure	£1,000	£1,500	£5,000	£7,000	£8,000	£4,000	£1,000
16								
17	Interest In Period	£0						
18	Cumulative Cost	£1,000						

In the first period, the total expenditure is £1,000 but there is no interest charged. The cumulative cost is therefore equal to the total expenditure.

E17=0
E18=E15

In the second period, interest is charged on the total cost accrued in the first period – but not on new costs in the second period. The cumulative cost is the previous cumulative cost, plus new expenditure and interest for current the period. So:

F17=((1+$E$10)^(1/12)–1)*E18
Interest Rate converted to a monthly figure times previous months cumulative cost.
F18=SUM(F15:F17)+E18
Total of expenditure and interest in month plus previous cumulative cost.

Copying these cells across will provide a running cumulative cost and interest calculation. The arrows on the illustration below help to show how the calculation works.

	D	E	F	G	H	I	J	K
10	Interest Rate	8%						
11								
12	Period	1	2	3	4	5	6	7
13	Date	Jan-04	Feb-04	Mar-04	Apr-04	May-04	Jun-04	Jul-04
14								
15	Total Expenditure	£1,000	£1,500	£5,000	£7,000	£8,000	£4,000	£1,000
16								
17	Interest In Period	£0	£6	£16	£48	£94	£146	£172
18	Cumulative Cost	£1,000	£2,506	£7,523	£14,571	£22,665	£26,811	£27,983
19								

End cost

Once the cashflow is complete you will need to refer to the final cumulative cost figure in your calculation of profit. Of course, you can easily do this by having, within the summary area, a fixed reference to the final cumulative cost amount.

However, if, for any reason, you amend, say, the build period and this affects the overall length of the cashflow you will have to remember to change this cell reference. It is easy to forget to do this.

Better if you use a cell reference that will automatically update itself. To do this you could create a calculation to work out how long the total development period is. You will probably be able to do this from inputs you are already using – for example, it might be the total of the scheme's planning, build and letting periods.

This sum gives the end period of the development. Incorporating this into a LOOKUP formula will select the appropriate figure as the constituent parts of the scheme's timing vary:

LOOKUP(End Period, Period Block, Cumulative Cost block)

Profit or land price?

The above assumes that you are entering all of the cost elements of the development in order to calculate the likely profit. Sometimes the reverse is true and the question is "If I need to generate a 20% profit, how much can I pay for the land?"

You could do this by varying the land price in your calculation until the profit reaches the required level. Alternatively, you can use the Excel spreadsheet's "Goal Seek" function to work out the land value.

To do this, choose the Tools/Goal Seek option, which will bring up this requester:

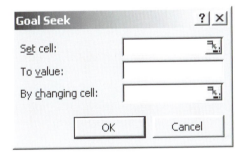

Fill this in – "Set cell" is the cell that contains your profit calculation, "To value" is the profit figure you require and "By changing cell" points to the land value input.

Excel will then calculate the appropriate land price without you having to use trial and error.

Chapter 14

Finishing Touches

A few ideas to help with your presentation

Hopefully you now have a broad range of ideas about how to set up cashflows and deal with a range of diverse situations. One area that we have not really touched on however, as it does not impact on the mechanics of the cashflow, is presentation.

Cashflows are complex things and contain a large amount of information, so making them easy to understand and clearly conveying your message is important – and this is a challenge more often than not.

Below are some ideas that may assist you in making your cashflows as easy to understand as you can.

Quarter day converters

There may be times when you prefer to use a tenancy schedule that shows the actual lease dates for the tenancies that you are generating cashflows for – but you still want to base your cashflow on the nearest quarter days. Alternatively, you may simply want to automate the process of rounding the dates to the nearest quarter day.

One way to carry out this task is to make use of the quarter days that you generate for your main cashflow. The actual dates can then be compared against this reference block to see whether the quarter day before or after the actual date is the closer.

In the following you might want to have your real tenancy schedule on the first page of your cashflow and these calculations on the second. Alternatively, you might simply hide the rounded dates once the cashflow is complete. Either way the techniques are very similar.

As an example, set up inputs as shown. Generate a block of quarter days to column BI or thereabouts (as you must make sure that it runs beyond the latest date you are likely to need to check).

	B	C	D	E	F	G	H	I
2		Original Dates			Quarter Days			
3	Start Date	Review	Expiry	Start Date	Review	Expiry	25-Dec-00	25-Mar-01
4								
5	04/03/2001	04/03/2006	03/03/2011					
6	01/08/2003		31/07/2008					

To compare the dates to the quarter days we will use HLOOKUP. First, you need to work out what the preceding quarter day is, you can use:

E5=HLOOKUP(B5,$H$3:$BI$3,1)
Search across the quarter day block for the date in B5 and return a result from the same row.

As the exact date does not appear in the block the formula will return the highest result before the date in B5 – i.e. the quarter day immediately before the date.

To get the quarter day following the date requires a bit of fiddling – and an acceptance of the fact that we do not actually need the following quarter day in all cases.

To get HLOOKUP to return the date of the next quarter you need to increase your date by just enough to make sure that the adjusted date that you are searching for falls after the next quarter day. Also, you know that if the date you are trying to round to a quarter day occurs in the first half of a quarter then the closest rounded date must be the preceding quarter day.

Traditional quarters are of varying length – between 87 and 97 days. So, if we add 50 days onto the date we are searching for then if it occurs in the second half of the month we can be sure that the next quarter day will be returned, if it is in the earlier part then the preceding quarter will be returned – but as we know that is the closest anyway it does not matter. So, to get the following quarter day you could replace the formula in E5 with:

E5=HLOOKUP(B5+50,$H$3:$BI$3,1)
Search across the quarter day block for the date in B5 plus 50 days and return a result from the same row.

Now that we know how to generate the quarters before and after the actual date (where needed) we can construct a formula to pick which is closest by comparing the number of days between the actual date and the quarters before and after:

E5=IF(B5-HLOOKUP(B5,$H$3:$BI$3,1)<HLOOKUP(B5+50,$H$3:$BI$3,1)- B5,HLOOKUP(B5,$H$3:$BI$3,1),HLOOKUP (B5+50,$H$3:$BI$3,1))
IF the actual date minus the previous quarter day is less than the following quarter day minus the actual date, then insert the quarter day before, otherwise insert the quarter day after

Having done that, the only refinement is to allow for blanks in your table of actual dates (for instance if there is no rent review in a particular lease). At the moment, the formula will generate an error if it tries to find a blank cell so, to avoid this, wrap the whole formula in:

=IF(B5= "", "",Previous Formula)
IF actual date cell is blank, return a blank, otherwise go through the previous formula.

So it will look like:

E5=IF(B5= "", "",IF(B5-HLOOKUP(B5,$H$3:$BI$3,1)<HLOOKUP (B5+50,$H$3:$BI$3 ,1)-B5,HLOOKUP(B5,$H$3:$BI$3,1), HLOOKUP(B5+50,$H$3:$BI$3,1)))

Which results in:

	B	C	D	E	F	G	H	I
2		Original Dates			Quarter Days			
3	Start Date	Review	Expiry	Start Date	Review	Expiry	25-Dec-00	25-Mar-01
4								
5	4-Mar-01	4-Mar-06	3-Mar-11	25-Mar-01	25-Mar-06	25-Mar-11		
6	1-Aug-03		31-Jul-08	24-Jun-03		24-Jun-08		

The above is quite tricky but is very easily adapted for different quarters dates or cashflow frequencies – monthly for example – as you will likely already have the appropriate date block generated and may simply have to adjust the number of days added onto your date (the "50" in the above example).

As noted earlier, this is only one way to approach the task. If you are looking to convert the dates but do not have a data block of quarter days to compare your input dates against you could start by working out the dates that mark the midpoints of the traditional quarters (8 February, 9 May, 11 August and 11 November).

Then you could use a series of IF statements to compare the input date with the midpoints. If your input date is in A1, start by checking against the first midpoint – 8 February:

=IF(A1<DATE(YEAR(A1),2,8),DATE(YEAR(A1)-1,12,25),...

If input date < 8 February in same year then nearest quarter is 25 December in previous year...

Then, simply keep adding on IF statements to check the remaining midpoints:

=IF(A1<DATE(YEAR(A1),2,8),DATE(YEAR(A1)-1,12,25), IF(A1<DATE(YEAR(A1),5,9), DATE(YEAR(A1),3,25), IF(A1<DATE(YEAR(A1),8,11),DATE(YEAR(A1),6,24), IF(A1< DATE(YEAR(A1),11,11),DATE(YEAR(A1),9,29)...

If Input Date < 8 February in same year then nearest quarter is 25 December in previous year, IF Input Date < 9 May then nearest quarter is 25 March, IF Input Date < 11 August then nearest quarter is 24 June, IF Input Date < 11 November then nearest quarter is 29 September...

Finally, if the date is after 11 November, then the nearest quarter must be 25 December in the same year, so insert this as false element of the last IF statement:

=IF(A1<DATE(YEAR(A1),2,8),DATE(YEAR(A1)-1,12,25), IF(A1<DATE(YEAR(A1),5,9), DATE(YEAR(A1),3,25), IF(A1<DATE(YEAR(A1),8,11),DATE(YEAR(A1),6,24), IF(A1 <DATE(YEAR(A1),11,11),DATE(YEAR(A1),9,29), DATE(YEAR(A1),12,25)))))

Both of these two methods will produce the same results for the traditional quarter days – but if you were rounding to the nearest month, for example, you would probably prefer the former approach.

Joining text and numbers

When summarising the results of your cashflow, there may be times when it would be helpful to combine the results of cells with text.

Perhaps a line such as "Based on an ERV of £10 psf, the IRR is 12%" would help clarify your results. Of course, it would be handy if this summary would update itself as results change, rather than you having to remember to change the figures.

On other occasions, you may want to include the units that a figure refers to within the cell:

Term	ERV		Void
25 years	£100,000	£10.00 psf	3 quarters

Rather than within the column heading:

Term	ERV		Void
(Years)	Total	Psf	(Quarters)
25	£100,000	£10.00	3

And yet still be able to use the entries as numbers. These types of formatting can easily be done in Excel.

Joining strings

The Excel function CONCATENATE enables us to join several pieces of information together – be they text, numbers or cell references. The function syntax is:

=CONCATENATE(Item 1, Item 2, Item 3…)

The main drawback with this is that formatting of the cells linked into the function is limited. Using the example from above, B4 in the example below contains:

B4=CONCATENATE("Based on an ERV of £",C2," psf, the IRR is ",C3*100,"%")

You can see how the formatting is incorporated – the "£" is manually input and the percentage is converted multiplied by 100 and placed in front of a "%" symbol – with the following result:

	B	C	D	E
2	ERV	£10		
3	IRR	12%		
4	Based on an ERV of £10 psf, the IRR is 12%			

Alternatively the "&" symbol can be used to join strings in a similar fashion. This has exactly the same result as using the CONCATENATE function:

B4="Based on an ERV of £"&C2&" psf, the IRR is " &C3*100&"%"

The "&" denotes where different strings are added together.

Custom formatting

To incorporate text with figures, as in the second of the above examples, you will need to create a new style. You will probably be familiar with formatting cells using the styles box:

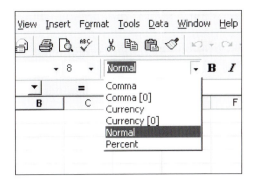

However, you can create your own formatting styles. Selecting the Format/Style menu option brings up:

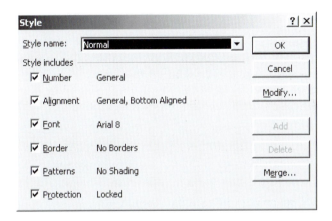

As an example, click in the "Style name:" box, type "Years" and then click on "Modify", to get:

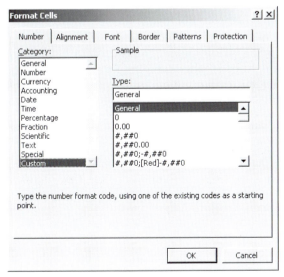

Selecting "Custom" will bring up the list as shown above. The Excel help file contains detailed information about the many different types of custom formatting available – but you can do a lot with just a few basic formats and some text:

displays significant digits only and ignores insignificant zeros.
#.## displays 123.00 as 123.
0 forces insignificant zeros to be included
0.00 displays 123.00 as 123.00
£ includes a currency symbol
£#.00 displays 123 as £123.00
" " can be used to incorporate text
£#.00" psf" displays 123 as £123.00 psf

So, as you will gather from the last formatting example, if you type the following over the word "General" in the "Type:" box:

#" years"

Then, once you have selected OK twice there will be a "Years" style in the styles box ready for you to format cells with. In the example below, the Term contains only the number 25 but is formatted with the Years style – which means that referring to this cell in formulae will refer to the number 25.

Term	ERV		Void
25 years	£100,000	£10.00 psf	3 quarters

The other styles are created in a similar fashion.

Selecting options

In some spreadsheets we have made use of text-based switches – such as using a "y" or "n" to select whether or not to include downwards rent reviews in a formula.
   You may also have need to switch between different scenarios – for example pessimistic, realistic and optimistic growth forecasts – where changing several inputs by hand might be laborious, especially if you need to do so frequently.
   Both of these situations can sometimes benefit from using the controls that can be embedded within a spreadsheet:

These are initially quite difficult to deal with – and may not seem worth the effort – but once you have used them a couple of time incorporating them becomes relatively straightforward.

Check boxes

The control pictured on the previous page does no more than insert a TRUE or FALSE result in a nominated spreadsheet cell. Your other formulae are then linked to this cell and react to its changes.

First you will need to access the "Control Toolbox" – selecting View/Toolbars and choosing Control Toolbox will bring up:

When you want to edit a control you will need to be in "Design Mode" – activated by clicking on the far left icon. The types of control available are the fourth to fourteenth icons.

To set up a check box, select the fourth icon and click again on the spreadsheet (within cell B2) – a plain box will appear labelled "Check Box 1".

If you now use the second icon, Properties, you will see:

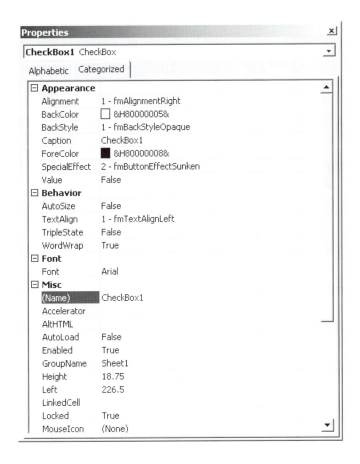

Select the "Categorised" tab. You can experiment with the options under appearance. Selecting "Back Colour" and choosing something other than white will help the button show up and the "Caption" option will allow you to change the displayed text.

Towards the middle of the "Misc" grouping you will see "Linked Cell". This is the address that the control will place its result in. Put C5 in here.

To test the button you must leave design mode – click the design mode button again to do this. As you click the check box, the state will change:

	B	C	D	E
2				
3	☑ CheckBox1			
4				
5		TRUE		
6				

	B	C	D	E
2				
3	☐ CheckBox1			
4				
5		FALSE		
6				

In practice, you might choose to hide the linked cell in an unused part of the sheet or on another sheet altogether – the format for this is "Sheet Name! Cell Reference", e.g. putting Sheet2!C5 in the "Linked Cell" box will make the result appear in C5 on Sheet 2.

List boxes

Considering the example of variable growth scenarios, we will look at how formulae can interact with the controls.

The check box above allowed us to choose whether a single item was TRUE or FALSE. List boxes extend this capability to allow the selection of one of several options.

Inserting a list box is done in much the same way as the check box. This time, select the eighth icon within the control toolbox and click once in C2. Once you have set up the headings, as shown, in row 9, you will have something that looks like this:

	B	C	D	E	F	G
2						
3						
4						
5						
6						
7						
8						
9	Growth Rates		2004	2005	2006	2007

The inputs that we will use for this example will be hidden on another page, so on Sheet 2, input the following:

	B	C	D	E	F	G
2	**Growth Rates**		**2004**	**2005**	**2006**	**2007**
3	Optimistic		5%	6%	7%	8%
4	Realistic		3%	4%	5%	6%
5	Pessimistic		1%	2%	3%	4%
6						
7	Selected					

In order to get the options into the list box you will need to revisit the properties box. While in design mode, select the list box and use the Properties icon (or right click on the list box and use same option on the menu that appears). You will get the following list:

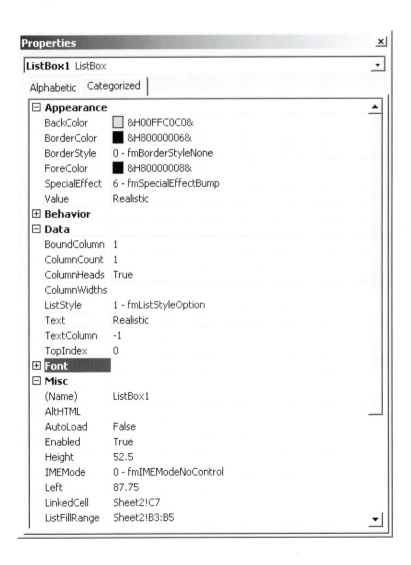

In this example, the Back Colour has been changed from white to make the control easier to see and the List Style has been changed to "1 – fm List Style Option" so that the list box incorporates option buttons in addition to text.

Towards the bottom of the list there is an input labelled "List Fill Range" which should contain the block of option labels that you want to incorporate. In this case it is "Sheet2!B3:B5". Above this is the "Linked Cell" input which should be "Sheet2!C7".

Finally, the "Column Heads" option within the Data inputs is set to "True" – this incorporates the "Growth Rates" heading in B2, immediately above the List Fill Range, as a title. This is the result:

	B	C	D	E	F	G
2						
3			Growth Rates			
4			● Optimistic			
5			○ Realistic			
6			○ Pessimistic			
7						
8						
9	Growth Rates		2004	2005	2006	2007

Now, leave design mode and choose an option. Whatever you choose, will be reflected in the "Selected" cell, C7, on the second sheet.

	B	C	D	E	F	G
2	Growth Rates		2004	2005	2006	2007
3	Optimistic		5%	6%	7%	8%
4	Realistic		3%	4%	5%	6%
5	Pessimistic		1%	2%	3%	4%
6						
7	Selected	Optimistic				

We can now use this within our formulae. To select the appropriate set of growth rates we can use the SUMIF function:

D7=SUMIF($B$3:$B$5,$C$7,D3:D5)
Look down B3:B5 for "Optimistic" and return the corresponding answer from D3:D5.

Note that the search column and search value cells are fixed, but the results column will move as the formula is copied across. Having copied the formula, the following will result:

	B	C	D	E	F	G
2	Growth Rates		2004	2005	2006	2007
3	Optimistic		5%	6%	7%	8%
4	Realistic		3%	4%	5%	6%
5	Pessimistic		1%	2%	3%	4%
6						
7	Selected	Optimistic	5%	6%	7%	8%

Finally, all that remains is to copy the results into your first page – if you want to be able to see what figures are being used:

D10=Sheet2!D7

This formula is copied into columns E, F and G. All of your formulae can then reference the first sheet and, as you select different scenarios, you will be able to see the growth rates that are being applied.

	B	C	D	E	F	G
2						
3			Growth Rates			
4			● Optimistic			
5			○ Realistic			
6			○ Pessimistic			
7						
8						
9	Growth Rates		2004	2005	2006	2007
10			5%	6%	7%	8%

Appendix

Frequently Used Functions

IF Functions

The IF function evaluates a logical test and returns a result based on whether the answer to the test is true or false:

=IF(Logical Test, TRUE, FALSE)

The logical test element of the formula can either be a mathematical expression that can be evaluated as being either true or false, or a function that returns a true or false result, for example the logical function AND.

Within the test, the mathematical expression uses comparison operators such as:

Equals	=
Greater than	>
Less than	<
Greater than or equal to	>=
Less than or equal to	<=
Not equal to	<>

IF statements can also be "nested" – when they take the following form, or a combination of these. Up to seven IF statements can be nested:

=IF(Logical Test, IF(Logical Test, TRUE, FALSE), FALSE)
=IF(Logical Test, TRUE, IF(Logical Test, TRUE, FALSE))

A related function is SUMIF, which is used as follows:

=SUMIF(Range, Criteria, Sum Range)

This function picks out the items in the range cells that satisfy the given criteria and adds together all of the related figures in the sum range.

The criteria can be an absolute figure, text or an expression making use of the comparison operators – for example ">5", "<>33". These must be contained within quotation marks.

If you have trouble using SUMIF, there is a wizard to help you through this step-by-step. Selecting Tools/Wizard/Conditional Sum ... will bring up the following for you to work through (see over).

If you cannot find this menu option, select the Tools/Add-Ins ... menu item and make sure that there is a check mark next to "Conditional Sum Wizard".

Logical functions

The logical operators take up to 30 arguments, evaluate each to a true or false answer and then generate an overall result. The two most commonly used are AND and OR.

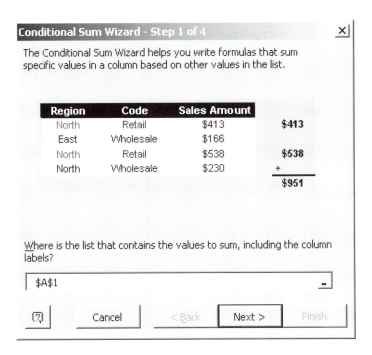

The Conditional Sum Wizard helps you write formulas that sum specific values in a column based on other values in the list.

Region	Code	Sales Amount	
North	Retail	$413	**$413**
East	Wholesale	$166	
North	Retail	$538	**$538**
North	Wholesale	$230	+
			$951

<u>W</u>here is the list that contains the values to sum, including the column labels?

$A$1

| ? | Cancel | < <u>B</u>ack | Next > | Finish |

=AND(Logical 1, Logical 2, ...up to 30 arguments)
=OR(Logical 1, Logical 2, ...up to 30 arguments)

The logical arguments are similar to the logical tests used in the IF statement and make use of the same comparison operators.

The AND function returns a TRUE result if all of the logical arguments within it are TRUE, otherwise the result is FALSE. OR returns a TRUE result if any of the logical arguments are TRUE, only if all of the results are FALSE will it return a FALSE answer.

Lookup functions

The lookup functions all search a range for a value and then return a corresponding result.

HLOOKUP and VLOOKUP are very similar:

=HLOOKUP(Value, Range, Row, Range Lookup)
=VLOOKUP(Value, Range, Row, Range Lookup)

Considering the HLOOKUP form:

Value is the item to be searched for and this can be a number or text. The search will be performed on the first row of the range.

Range (referred to in the Excel Help text as "table array") is the block to search and must include all of the rows that you wish to search down.

Row is the number of rows to look down once the Value is found. 1 will return a value from the same row that the search is performed on. If you try and return a value from a row that is not contained within the Range block, then a "#Ref!" error will be generated.

Range Lookup is used to decide whether you want an exact or approximate match to be found. If this is set to FALSE then only an exact match will be used. If an exact match does not exist then a "#N/A" error will be returned. If this is set to TRUE, or if it is omitted, then an approximate match is used. In this case, if there is no exact match then the next highest result in the range will be used. For this reason, if the TRUE option is to be used then the values in the first row of the Range must be in ascending order.

VLOOKUP works in exactly the same way, save that it searches down a column and then across rather than along a row and then down.

The plain LOOKUP function works slightly differently.

=LOOKUP(Value, Lookup Range, Results Range)
Value is the item to be searched for and this can be a number or text.
Lookup Range is a single row or column that is to be searched.
Results Range is also a single row or column of the same size from which the corresponding result is selected.

This function does not have an option to select an exact match. If no exact match exists then the result corresponding to the next highest match is returned.

Date functions

By default, Excel stores dates as sequential numbers beginning with 1, which represents 1 January 1900, and these are then formatted to display dates. Dates can be entered either by directly typing them – 25–Mar–03 and 25/3/03 will both work – or through a formula:

=DATE(Year, Month, Day)
For example =Date(2003,3,25)

Working from the opposite direction, three functions can be used to extract year, month and day elements from a date.

=YEAR(Cell) returns the year represented by the value in the selected Cell.
=MONTH(Cell) and =DAY(Cell) both work in the same way.

These can be used in conjunction with the Date formula to perform a variety of complex date calculations.

In order to add a number of months to a date without adjusting the day figure, EDATE can be used:

=EDATE(Start Date, Months)
Adds the specified number of months onto the Start Date

Financial functions

There are a wide number of financial functions within Excel, some of which can only be accessed if the Analysis Toolpak has been added in (see chapter 3 for details). The loan functions used in this book are as follows.

To calculate the periodic payment on a loan:

=PMT(Rate, Number of Payments, Present Value, Future Value, Type)
Rate is the periodic interest rate for the loan.
Number of Payments is the total number of payments for the loan.
Present Value is the amount of the loan.
Future Value is the amount to which the loan is to be amortised to. If you are treating the loan amount as a positive number, this figure must be negative.
Type allows you to choose payments in advance (value is 1) or arrears (value is 0 or omitted).

The periodic payment comprises two elements – the interest and capital repaid in each period. These figures can be determined individually using two very similar functions:

=IPMT(Rate, Period, Number of Payments, Present Value, Future Value, Type)
=PPMT(Rate, Period, Number of Payments, Present Value, Future Value, Type)
The constituents of this function are the same as for the PMT function save that a Period input (i.e. what period are you looking at) is required as the ratio between the interest and capital repaid varies over time.

The final loan formula used considers the cumulative amount of the principal of a loan repaid between two points in time:

=CUMPRINC(Rate, Number of Payments, Present Value, Start Period, End Period, Type)
Rate and Number of Payments are as used in the previous functions.
Present Value is the amount of capital to be repaid. As there is no Future Value input then this must equal the amount borrowed less the amount you wish to amortise to.
Start and End Period are the points between which you want to calculate the amount of capital repaid. These can be the same if you want the amount in a single period.
Type is, as usual, used for timing of the payment (0 for payment at the end of the period, 1 for at the start). However, in this function, it must be included (in previous functions it could be omitted).

The last two financial functions used are for calculating net present values and internal rates of return. The NPV function is:

=NPV(rate,value1,value2, ...)
Rate is the discount rate on the same periodic basis as your cashflow (i.e. annual, quarterly etc).
Value can either be individual figures or a range.

With the NPV calculation it is important to remember that by default it will treat payments as if received in arrears. If payments are made in advance then the first payment should be left out of the calculation and then added in separately.
The IRR function takes the form:

=IRR(Value Range, Guess)
Value Range covers the cashflow that you wish to calculate the IRR for and must contain at least one positive and one negative value.
Guess provides the IRR with a start figure from which to begin its iterative calculation. If left blank then 10% will be assumed.

Miscellaneous

There are two functions for determining the maximum and minimum values in a list:

=MAX(Value 1, Value 2, up to Value 30)
=MIN(Value 1, Value 2, up to Value 30)
The Values can be numbers or ranges.

Finally, the MOD function can be used to determine the remainder left after a division is carried out:

=MOD(Number, Divisor)
Number is the number for which you wish to find the remainder
Divisor is the number that you want to divide by.